Ergebnisse der Mathematik
und ihrer Grenzgebiete   Band 86

A.M. Olevskiĭ

Fourier Series
with Respect to General
Orthogonal Systems

Translated from the Russian
by B.P. Marshall and H.J. Christoffers

Springer-Verlag
Berlin Heidelberg New York 1975

Professor Alexander Olevskiĭ
Moscow Institute of Electronic Machine Construction
Faculty of Applied Mathematics

AMS Subject Classification (1970): 42-02, 42 A 52, 42 A 56, 42 A 60, 42 A 62

ISBN-13: 978-3-642-66058-0   e-ISBN-13: 978-3-642-66056-6
DOI: 10.1007/978-3-642-66056-6

Library of Congress Cataloging in Publication Data. Olevskiĭ, A. M. 1939—. Fourier series with respect to general orthogonal systems. (Ergebnisse der Mathematik und ihrer Grenzgebiete; Bd. 86.) Translation of Riady Fur'e po obshchim ortogonal'nym sistemam. Bibliography: p. 1. Functions, Orthogonal. 2. Series, Orthogonal. 3. Fourier series. I. Title. II. Series. QA404.5.04313. 515'.55. 74-32297

In memory of my mother

FRIDA OLEVSKAYA

Preface

The fundamental problem of the theory of Fourier series consists of the investigation of the connections between the metric properties of the function expanded, the behavior of its Fourier coefficients $\{c_n\}$ with respect to an orthonormal system of functions $\{\phi_n\}$, and the convergence of the series $\sum c_n \phi_n(x)$.

This problem has many different aspects. First of all, should we examine these questions for arbitrary orthogonal systems, for those systems that satisfy certain general assumptions, or for a particular system $\{\phi_n\}$? Furthermore, the convergence of a series can be interpreted in different ways—in the classical sense, almost everywhere, with respect to a certain metric, in the sense of some summability process.

The problems mentioned above arose in connection with the application of the Fourier method to problems of mathematical physics. Subsequently they turned out to be very pithy in their own right; from their investigation arose ideas and methods that exerted considerable influence on other areas of mathematics—functional analysis, the theory of analytic functions, probability theory, etc.

This applies first of all to the theory of trigonometric Fourier-Lebesgue series, which is most highly developed. This extensive and deep theory is the subject of the famous monographs of A. Zygmund [171] and N. K. Bary [10].

Investigations of general orthogonal systems are contained in the monograph of S. Kaczmarz and H. Steinhaus [55]. Later work is presented in the appendix to the Russian translation [47] and in the book by G. Alexits [1].

These monographs contain, in particular, accounts of the classical results of A. N. Kolmogorov, D. E. Menshov, A. Haar, J. Marcinkiewicz and other authors who have made basic contributions to the development of the theory and have had a considerable influence on the formation of the problems.

The last several years have been a period of intensive development in the theory of Fourier series. Fundamental progress in trigonometric series has been achieved by the work of L. Carleson [18], who gave a positive solution to the famous problem of Lusin concerning the convergence almost everywhere of Fourier series in L^2.

Advances have also been made in the theory of Fourier series with respect to general orthogonal systems. In particular, it was discovered that several results which had seemed to be connected specifically to the trigonometric system have in fact a considerably more general nature and are determined by such properties of orthonormal systems as completeness or uniform boundedness.

Thus it was shown that the classical fact of the divergence of the Fourier series of a continuous function (the growth of the Lebesgue constants) holds for any bounded orthonormal system (Chap. I).

Further it was discovered that the essential role of the method of ordering the trigonometric system, in the problem of the convergence almost everywhere, is due to the completeness of the system. Meanwhile the special role of the Haar system in the class of all complete systems was established; a method was developed (Chap. III) permitting in a number of cases a reduction of the problem for an arbitrary complete system to the same problem for the Haar system.

For general orthogonal systems the convergence almost everywhere of a particular L^2-series has been proved to occur for an appropriate rearrangement of the terms. Definitive results have been obtained on coefficient conditions for convergence (Chap. II).

At the same time, in a number of questions it turned out that there is a variety of different possibilities, and examples have been found of systems that have essentially different and sometimes surprising properties (Chap. IV).

This monograph is devoted to these questions; it is based primarily on the investigations of the last fifteen years concerning Fourier series with respect to general orthogonal systems. Results involving specific systems are examined only to the extent that they shed light on the problems of the general theory. We will not touch at all upon the investigation of multiple Fourier series and spectral expansions, or upon multiplicative systems and other special classes of orthonormal systems.

The bibliography contains only the works cited. The works [7, 39, 54, 84, 138, 163] contain a survey of important results that are related to our theme but are either not given in sufficient detail or not mentioned at all in this book.

We particularly mention the survey by P. L. Ulyanov [158], in which a number of the problems examined below are formulated.

The fundamental results are given with proofs; however, the author has tried to avoid letting the technical details encumber the presentation. A number of the results of Chap. I and III were formerly only announced in [100, 101, 106] and are now for the first time set forth in detail.

Part of the contents of this book formed the subject of lectures given by the author at the University of Szeged (Hungary) in the winter of 1970/71. The initiative for the writing of the present volume came from Professor B. Sz.-Nagy, to whom the author expresses deep gratitude.

Contents

Terminology. Preliminary Information

(for more detail see [1,8,10,55])

We shall examine orthonormal systems (ONS) in $L^2(X)$, where X is the interval $[a,b]$ with Lebesgue measure; however, many results extend more or less automatically to any compact space with a finite Borel measure, or in general to any measure space. [This is true, for example, of the inequality in Chap. I.]

As a rule we shall have the real case in mind. $\|f\|_p$ will denote the norm in the space $L^p(X), 1 \le p \le \infty$ (L^∞ is the space of essentially bounded measurable functions). μE will denote the measure of the set $E \subset X$; $\chi(E;x)$ will be the characteristic function of the set E.

If an orthogonal series (that is, a series with respect to an ONS)

$$\sum c_n \phi_n \tag{1}$$

converges to a function f in the L^p-metric, and if

$$\phi_n \in L^q, \quad q = \frac{p}{p-1}, \tag{2}$$

then the coefficients are uniquely defined by

$$c_n = (f, \phi_n) \equiv \int_X f \phi_n \, dx. \tag{3}$$

With condition (2), to each function $f \in L^p$ there can be associated by means of equation (3) a sequence $\{c_n\}$ (the *Fourier coefficients* of the function f with respect to the system $\{\phi_n\}$), and consequently also a *Fourier series* (1). This series need not converge.

The system $\phi = \{\phi_n\}$ is said to be *closed* in the Banach space B if the set P_ϕ of all linear combinations $\sum_1^n \alpha_k \phi_k \, (n=1,2,...)$ is everywhere dense in B. We shall use the phrase "the system is closed in C" even in the case where ϕ_n are piecewise continuous, meaning that every continuous function can be uniformly approximated by linear combinations of the system ϕ.

The system ϕ will be called *complete* in $L^p, 1 \le p \le \infty$, if condition (2) is satisfied and if $L^p \ni f \ne 0$ implies $\sum |c_n(f)| > 0$. Completeness of a system in $L^p \, (p>1)$ is equivalent to its being closed in L^q. The term "complete system" (without specifying the space) will mean completeness in L^2. If a system is closed in C then it is complete in every L^p.

The system ϕ is called a *basis* in the separable space B if every vector $f \in B$ expands in a unique way as a convergent series $\sum c_k \phi_k$. In this case there exists a system of functionals $\psi_k \in B^*$ dual to $\{\phi_n\}$; that is, $\psi_k(\phi_n) = \delta_{k,n}$ (δ is the Kronecker delta) and $c_k = \psi_k(f)$.

In order that a system of vectors ϕ, closed in B, form a basis for this space it is necessary and sufficient that the following quantity be finite (the "Banach constant"):

$$K(\phi) = \sup_{\substack{p = \sum_1^n d_k \phi_k \\ \|p\| \leq 1}} \max_{1 \leq l \leq n} \left\| \sum_{k=1}^l d_k \phi_k \right\|.$$

Two systems of vectors will be called *equivalent* if each vector of one system can be expressed as a linear combination of elements of the other system.

H^ω will denote the class of functions f satisfying the condition $\omega(\delta; f) = O(\omega(\delta))$, where $\omega(\delta)$ is a given modulus of continuity and $\omega(\delta; f)$ is the modulus of continuity of the function f. H_p^ω will be the corresponding class in the space L^p. In particular (in the case $\omega(\delta) = \delta^\alpha, 0 < \alpha \leq 1$) H^α (H_p^α) is a Hölder class.

V is the space of all functions of bounded variation on $[a,b]$,

$$\|f\|_V = \operatorname*{Var}_{[a,b]} f.$$

For the convenience of the reader we list the definition of some classical ONS.

The Haar system. Let $\chi_0^{(0)} \equiv 1$ and

$$\chi_k^{(j)}(x) = \begin{cases} \sqrt{2^k}, & x \in \left(\dfrac{j-1}{2^k}, \dfrac{j-\frac{1}{2}}{2^k} \right) \\[2ex] -\sqrt{2^k}, & x \in \left(\dfrac{j-\frac{1}{2}}{2^k}, \dfrac{j}{2^k} \right) \quad 1 \leq j \leq 2^k; k = 0,1,\ldots \\[2ex] 0, & x \notin \left[\dfrac{j-1}{2^k}, \dfrac{j}{2^k} \right] \end{cases}$$

At the interior points of discontinuity define $\chi_k^{(j)}$ to be equal to the average of the limits on either side; at the endpoints of the interval $[0,1]$ define the function to be the limits from the interior (such a definition of a piecewise continuous function at the points of discontinuity is called *regular*).

There is a natural ordering of the Haar system:

$$\chi_1 = \chi_0^{(0)}; \ \chi_n = \chi_k^{(j)} \quad (n = 2^k + j, \ 1 \leq j \leq 2^k).$$

This system is orthonormal on $[0,1]$ and closed in C.

The Rademacher system. Let

$$r_k(x) = (-1)^{j-1}, \quad x \in \left(\frac{j-1}{2^k}, \frac{j}{2^k} \right), \quad 1 \leq j \leq 2^k, k = 1,2,\ldots$$

and extend regularly at the points of discontinuity.

The system $\{r_k\}$ is orthonormal on $[0,1]$, but is far from complete. It is one of the simplest examples of a sequence of stochastically independent functions.

The following inequality of Khinchin is true:

$$C_p \sqrt{\sum_{k=1}^{n} \alpha_k^2} \leq \left\| \sum_{k=1}^{n} \alpha_k r_k \right\|_p \leq C_p' \sqrt{\sum_{k=1}^{n} \alpha_k^2},$$

$$C_p, C_p' > 0, 1 \leq p < \infty, \quad n = 1, 2, \ldots.$$

The Walsh system. Let $w_1(x) \equiv 1$; further, if $n - 1 = \sum_{k=0}^{s} \varepsilon_k 2^k$ $(\varepsilon_k = 0, 1)$ is the binary expansion of the number $n - 1$, then define

$$w_n(x) = \prod_{k; \varepsilon_k = 1} r_{k+1}(x).$$

$\{w_n\}$ is the Walsh system with the Paley ordering. It is orthonormal and complete on $[0, 1]$ and together with the trigonometric system it belongs to the class of multiplicative systems. This ensures a similarity of the properties of these systems in a number of respects.

The subspace $X_\nu(W_\nu)$ generated by the functions $\{\chi_n\}$, $1 \leq n \leq 2^\nu$ (respectively, $\{w_n\}$, $1 \leq n \leq 2^\nu$), coincides with the set of all step functions constant on each interval $\left(\frac{j-1}{2^\nu}, \frac{j}{2^\nu} \right)$ (and regular at the points of discontinuity).

Chapter I. Convergence of Fourier Series in the Classical Sense. Lebesgue Functions of Bounded Systems

The classical aspect of the theory of Fourier series consists of the expansion of a given periodic function f as a trigonometric series $\sum c_n e^{inx}$ converging to f uniformly or at each point. Numerous results show that if f possesses a certain smoothness, for example, if it belongs to the Hölder class $H^\alpha, \alpha > 0$, then it expands as a uniformly convergent trigonometric Fourier series.

Just the condition of continuity on the function is not sufficient: as early as the end of the last century du Bois-Reymond discovered that the Fourier series of a continuous function can diverge at some points.

In this connection arose the problem of the existence of a system of functions $\phi = \{\phi_n\}$ orthonormal on $X = [a,b]$ that has the property that each function $f \in C(X)$ expands as a uniformly convergent Fourier series

$$\sum c_n \phi_n(x). \tag{1}$$

This problem was investigated in the doctoral thesis of A. Haar [48]. The system $\{\chi_n\}$ that he constructed (see above) has the property mentioned.

It is true that these functions are only piecewise continuous, but this is not essential. Shortly after the work of Haar, Franklin gave a construction of an ONS consisting of continuous functions and having the same property. The Franklin system (see § 5) is the first example of an orthonormal basis in the space C.

The question of the convergence of the Fourier series (1) leads to the concept of Lebesgue functions. The partial sum $S_n(f; x) = \sum_{k=1}^{n} c_k(f) \phi_k(x)$ represents a continuous linear functional on the space C. It can be written in the form

$$S_n(f; x) = \int_X \mathscr{D}(x,t) f(t) dt,$$

where $\mathscr{D}_n(x,t) = \sum_{k=1}^{n} \phi_k(x) \phi_k(t)$ is the Dirichlet kernel. The norm of this functional.

$$L_n^\phi(x) = \int_X \left| \sum_{k=1}^{n} \phi_k(x) \phi_k(t) \right| dt,$$

is called a *Lebesgue function* of the ONS ϕ. According to the Banach-Steinhaus theorem, in order that a given point x satisfy the equality

$$f(x) = \sum_{n=1}^{\infty} c_n(f) \phi_n(x) \quad (\forall f \in C)$$

it is necessary that the condition

$$L_n(x) = O_x(1)$$

be fulfilled. Analogously, for the uniform convergence of the Fourier series of any continuous function it is necessary that

$$\sup_{x \in X} L_n(x) = O(1).$$

With the assumption that the system be closed in the space C, these conditions are also sufficient.

For the trigonometric system the Lebesgue functions are not difficult to calculate:

$$L_n(x) \equiv L_n \sim K \ln n \quad (K > 0) \tag{2}$$

and from this follows the theorem of du Bois-Reymond. For the Haar system we have instead the equality $L_n(x) \equiv 1 \; (\forall n)$.

On comparison of the trigonometric and Haar systems, the following difference comes to our attention: the trigonometric functions are uniformly bounded, whereas the Haar system is built from high narrow peaks.

Not long ago it was discovered that this difference plays an essential role in the questions we are examining. In [97] the author established that an orthonormal system $\{\phi_n\}$ cannot be uniformly bounded at the same time that its Lebesgue functions are. More precisely, if an ONS satisfies the condition

$$|\phi_n(x)| < M \tag{3}$$

then every x lying in some set of positive measure satisfies the condition

$$L_n(x) \neq O(1).$$

This shows that *it is impossible to construct a bounded ONS such that every continuous function has an everywhere convergent Fourier series expansion.* Thus the phenomenon of the local divergence of a Fourier series (the du Bois-Reymond theorem) is connected not specifically with the trigonometric system, but has a general nature—it arises with any ONS satisfying condition (3).

Further, it has been stated in [101] that the uniform boundedness of an ONS determines a fixed order of growth of the Lebesgue functions. Namely, such systems always satisfy the relation

$$L_n(x) \neq o(\ln n) \quad (x \in E, \mu E > 0).$$

The accuracy of this statement can be seen from (2).

These results and others connected with them form the content of the present chapter. They are based on a new method of estimating a lower bound in the metric of L for the partial sums of series of orthogonal functions (§ 1). § 2 is concerned with applications to the Fourier series of continuous functions. Some other applications of this method are contained in § 3.

In § 5 the stability properties of the classical process of orthogonalization are investigated, and applications to the construction of orthogonal bases in function spaces are indicated.

§ 1. The Fundamental Inequality

Let X be a space with measure μ, and let $\{\phi_k\}$ $(1 \leq k \leq n)$ be an orthonormal sequence of functions (real or complex) in the space $L^2(X)$. Assume

$$\max_{1 \leq k \leq n} \|\phi_k\|_\infty = M . \tag{1}$$

If $c = \{c_k\}$ $(1 \leq k \leq n)$ is a sequence of numbers. then by $\|c\|_\infty$ we will mean $\max |c_k|$. A fundamental role in this chapter is played by the following proposition.

Theorem 1. *The following inequality holds:*

$$\|c\|_\infty \max_{1 \leq k \leq n} \int_X \left| \sum_{v=1}^{k} c_v \phi_v(t) \right| d\mu \geq \frac{K}{M} \frac{\sum_{v=1}^{n} |c_v|^2}{n} \ln n , \tag{2}$$

where $K > 0$ is an absolute constant.

This result with applications (§ 2) was stated in [101].

Because of the homogeneity of inequality (2) and its invariance relative to the transformation $\hat{\mu} = \delta\mu$, $\hat{\phi}_n(x) = \dfrac{1}{\sqrt{\delta}} \phi_n(x), \delta > 0$, we can assume that

$$\|c\|_\infty = M = 1 . \tag{3}$$

Let $c = \{c_k\}$, $1 \leq k \leq n$, be a fixed sequence satisfying condition (3). By Δ we will mean any interval of natural numbers of the form $[v+1, v+p^r] \subset [1, n]$ $(r \geq 0$ is an integer, $p = 34)$. We introduce into consideration the following indices:

the density $\alpha(\Delta) = \dfrac{1}{|\Delta|} \sum_{k \in \Delta} |c_k|^2$, $\qquad |\Delta| = p^r$.

the content $\gamma(\Delta) = (r+1)\alpha(\Delta)$.

For each interval Δ $(r > 0)$, we define

$$\Delta^i = [v+(i-1)p^{r-1} + 1, v + i p^{r-1}] \quad (1 \leq i \leq p) . \tag{4}$$

Call Δ a normal interval if

$$\gamma(\Delta) \geq \gamma(\Delta^i) \quad (\forall i) \tag{5}$$

(when $r = 0$ then Δ is normal by definition).

The following proposition is an inductive lemma in the proof of the theorem.

Lemma. *Let f be a function such that*

$$\|f\|_\infty \leq p^{r+1}, \quad r > 2p , \tag{6}$$

and let $\Delta = [v+1, v+p^r]$ be a normal interval. Then there exist numbers l, k that satisfy the following conditions:
(i) the interval $\delta = [l+1, l+p^k]$ is normal, $v < l < l + p^k \leq v + p^r$;

(ii) $\lambda \int_{U_k^l} \left| f + \sum_{i=v+1}^{l} c_i \phi_i \right| d\mu \geq \gamma(\Delta) - \gamma(\delta)$,

where

$$U_k^l = \left\{ t : \left| f(t) + \sum_{i=v+1}^{l} c_i \phi_i(t) \right| > p^{k+1} \right\} \quad (k > 0); \quad U_0^l = X, \tag{7}$$

and λ is an absolute constant.

Since Δ is normal we have

$$\alpha(\Delta^i) \leq \frac{r+1}{r} \alpha(\Delta) \quad (1 \leq i \leq p). \tag{8}$$

Hence for each fixed j it follows that

$$\alpha(\Delta) = \frac{1}{p^r} \sum_{k \in \Delta} |c_k|^2 = \frac{1}{p} \sum_{i=1}^{p} \alpha(\Delta^i) = \frac{1}{p} \left[\alpha(\Delta^j) + \sum_{i \neq j} \alpha(\Delta^i) \right] \leq \frac{\alpha(\Delta^j)}{p} + \frac{p-1}{p} \frac{r+1}{r} \alpha(\Delta).$$

That is,

$$\alpha(\Delta^j) > \frac{r-p}{r} \alpha(\Delta) \quad (1 \leq j \leq p). \tag{9}$$

We shall define a nondecreasing sequence of numbers $\{v_s\}$ $(1 \leq s \leq r)$. If the inequality

$$\int_X |f|^2 dt > 9 p^{r-1} \alpha(\Delta) \tag{10}$$

is satisfied, then define $v_1 = v + p^{r-1}$; otherwise define $v_1 = v + (p-1) p^{r-1}$. The numbers $v_2 \leq v_3 \leq \cdots \leq v_r$ are chosen successively so that the intervals $\Delta_s = [v_s + 1, v_s + p^{r-s}]$ satisfy the relations

$$\Delta_s = \Delta_{s-1}^{i_s}, \quad \alpha(\Delta_s) \geq \frac{r-p}{r} \alpha(\Delta). \tag{11}$$

It is possible to do this, taking into account (9), since at each step for at least one value $i = i_s$ the inequality $\alpha(\Delta_{s-1}^{i_s}) \geq \alpha(\Delta_{s-1})$ is satisfied. Clearly,

$$v_r - v_s \leq p^{r-s} \quad (1 \leq s \leq r). \tag{12}$$

Define

$$f_s = f + \sum_{i=v+1}^{v_s} c_i \phi_i. \tag{13}$$

$$U_s = \{t : |f_s(t)| > p^{r-s+1}\} \quad (s < r), \quad U_r = X. \tag{14}$$

We note the following inequality:

$$\int_X |f_r(t)|^2 d\mu > p^{r-1} \alpha(\Delta). \tag{15}$$

Indeed, if (10) is satisfied, then taking into account (12) and (8), we have

$$\|f_r\|_2 = \left\| f + \sum_{i=v+1}^{v_r} c_i \phi_i \right\|_2$$

$$\geq \|f\|_2 - \left(\sum_{i=v+1}^{v_r} |c_i|^2 \right)^{1/2} > (9\,p^{r-1}\alpha(\Delta))^{1/2} - \left(\sum_{i=v+1}^{v+2\,p^{r-1}} |c_i|^2 \right)^{1/2}$$

$$= 3\left[p^{r-1}\alpha(\Delta) \right]^{1/2} - (p^{r-1}[\alpha(\Delta^1) + \alpha(\Delta^2)])^{1/2}$$

$$\geq \left[p^{r-1}\alpha(\Delta) \right]^{1/2} \left[3 - \left(2\,\frac{r+1}{r} \right)^{1/2} \right].$$

If inequality (10) is not satisfied, then we estimate the value analogously, keeping in mind (9) and (6):

$$\|f_r\|_2 \geq \left(\sum_{i=v+1}^{v_r} |c_i|^2 \right)^{1/2} - \|f\|_2 \geq \left(\sum_{i=v+1}^{v_1} |c_i|^2 \right)^{1/2} - \|f\|_2$$

$$= \left(\sum_{j=1}^{p-1} \sum_{i\in\Delta^j} |c_i|^2 \right)^{1/2} - \|f\|_2 \geq \left[p^{r-1}\alpha(\Delta) \right]^{1/2} \left[\left(\frac{(p-1)(r-p)}{r} \right)^{1/2} - 3 \right]$$

$$> \left[p^{r-1}\alpha(\Delta) \right]^{1/2} \left[\left(\frac{p-1}{2} \right)^{1/2} - 3 \right] > \left[p^{r-1}\alpha(\Delta) \right]^{1/2}.$$

Further, we notice the following inequality:

$$|f_r(t)| < 2|f_s(t)| \qquad (t \in U_s,\ 1 \leq s < r). \tag{16}$$

Indeed, using in order (13), (1), (3), (12), (14), we have

$$|f_r(t)| \leq |f_s(t)| + \left| \sum_{i=v_s+1}^{v_r} c_i \phi_i(t) \right| \leq |f_s(t)| + p^{r-s} < |f_s(t)| \left(1 + \frac{1}{p} \right).$$

Analogously, because of (6),

$$\|f_r\|_\infty < p^{r+2}. \tag{17}$$

Define

$$\max_{1 \leq s \leq r} \frac{1}{s} \int_{U_s} |f_s(t)|\, d\mu = \rho \tag{18}$$

and estimate $\|f_r\|_2$ from above. Let

$$G_s = \{t;\ p^{r-s+2} < |f_r(t)| \leq p^{r-s+3}\} \quad (1 \leq s < r), \quad G_r = \{t;\ |f_r(t)| \leq p^3\}.$$

Clearly, for $t \in G_s\,(s < r)$, we have

$$|f_s(t)| \geq |f_r(t)| - p^{r-s} > p^{r-s+2} - p^{r-s} > p^{r-s+1}.$$

That is, $G_s \subset U_s\ (1 \leq s \leq r)$. Therefore, because of (17), (16), and (18), we arrive at the estimate

$$\int_X |f_r|^2 d\mu = \sum_{s=1}^{r} \int_{G_s} |f_r|^2 d\mu \le \sum_{s=1}^{r} \sup_{t \in G_s} |f_r(t)| \int_{G_s} |f_r| d\mu$$

$$\le \sum_{s=1}^{r} p^{r-s+3} \int_{U_s} |f_r| d\mu \le 2 p^{r+3} \sum_{s=1}^{r} p^{-s} \int_{U_s} |f_s| d\mu$$

$$\le 2 p^{r+3} \rho \sum_{s=1}^{r} p^{-s} s < 2 p^{r+3} \rho.$$

Comparing this with (15), we conclude that $\rho > \dfrac{\alpha(\varDelta)}{2 p^4}$. Therefore, for some number $s = s_0$ we have the inequality

$$\int_{U_{s_0}} |f_{s_0}| d\mu > \frac{1}{2 p^4} \alpha(\varDelta) s_0. \tag{19}$$

Furthermore, it is easy to see from the definition of normality that there can be found a normal interval

$$\delta = [l+1, l+p^k] \subset \varDelta_{s_0}, \quad \gamma(\delta) \ge \gamma(\varDelta_{s_0}) \ge (r-s_0+1) \frac{r-p}{r} \alpha(\varDelta) \tag{20}$$

(the latter is a result of (11)). Let

$$f + \sum_{i=v+1}^{l} c_i \phi_i \equiv f_{s_0} + \sum_{i=v_{s_0}+1}^{l} c_i \phi_i = F. \tag{21}$$

We have the inclusion

$$U_{s_0} \subset U_k^l \tag{22}$$

(the latter set is defined by relation (7)). Indeed, if $k = r - s_0$, that is, $\delta = \varDelta_{s_0}$, then these two sets coincide. In the case $k < r - s_0$ we have, for $t \in U_{s_0}$,

$$|F(t)| \ge |f_{s_0}(t)| - (l - v_{s_0}) > p^{r-s_0+1} - p^{r-s_0} \ge p^{k+1}$$

whence follows (22). Further, for $t \in U_{s_0}$ the following inequality is true:

$$|F(t)| > |f_{s_0}(t)| - p^{r-s_0} > |f_{s_0}(t)| \left(1 - \frac{1}{p}\right).$$

Therefore, keeping in mind (22) and (19), we obtain

$$\int_{U_k^l} |F(t)| d\mu \ge \tfrac{1}{2} \int_{U_{s_0}} |f_{s_0}| d\mu > \frac{s_0}{4 p^4} \alpha(\varDelta). \tag{23}$$

Finally, letting $\lambda = p^6$, because of (20) and (22), we conclude that

$$\lambda \int_{U_k^l} |F| d\mu + \gamma(\delta) > \alpha(\varDelta) \left[3 p s_0 + (r - s_0 + 1) \frac{r-p}{r} \right]$$

$$> (r+1) \alpha(\varDelta).$$

This completes the proof of the lemma.

We now turn to the proof of inequality (2). Let $r_0 = [\log_p n]$. Clearly we can define an interval $\Delta_0 = [v_0 + 1, v_0 + p^{r_0}] \subset [1, n]$ that satisfies the inequality

$$\alpha(\Delta_0) \geq \frac{1}{2n} \sum_{k=1}^{n} |c_k|^2 . \tag{24}$$

and can then choose a normal interval $\delta_0 = [l_0 + 1, l_0 + p^{k_0}] \subset \Delta_0$ such that

$$\gamma(\delta_0) \geq \gamma(\Delta_0) . \tag{25}$$

Henceforth it will be assumed that $k_0 > 2p$; otherwise the proof is completed by calculation (32), see below, where $q = 0$.

We shall make use of the following notation (where the numbers l_s and k_s are defined below):

$$F_0 \equiv 0 ; \quad F_s = \sum_{i = l_0 + 1}^{l_s} c_i \phi_i \ (s > 0); \quad E_s = \{t; |F_s(t)| > p^{k_s + 1}\} \ (k_s > 0);$$

$$f_s(t) = F_{s-1}(t) \chi(\complement E_{s-1}; t) + \sum_{i = l_{s-1} + 1}^{l_s} c_i \phi_i(t) \tag{26}$$

$$U_s = \{t; |f_s(t)| > p^{k_s + 1}\} \ (k_s > 0); \quad E_s = U_s = X \ (k_s = 0);$$

and further

$$I_s = \int_{E_s} |F_s(t)| d\mu ; \quad J_s = \int_{U_s} |f_s(t)| d\mu \tag{27}$$
$$(\complement E = X \setminus E) .$$

The numbers $l_s, k_s, 1 \leq s \leq q$ are chosen so that
  (a) $l_{s-1} < l_s, k_{s-1} > k_s$;
  (b) the interval $\delta_s = [l_s + 1, l_s + p^{k_s}]$ is normal, $\delta_s \subset \delta_{s-1}$;
  (c) $\lambda J_s \geq \gamma(\delta_{s-1}) - \gamma(\delta_s)$;
  (d) $k_q \leq 2p$.

This is done by induction in the following way. Let the numbers $l_s, k_s, 0 \leq s \leq m$ be already defined so that they satisfy conditions (a)–(c). If $k_m \leq 2p$, then let $q = m$ and the process is finished. If $k_m > 2p$, then let $v = l_m, r = k_m, f = F_m \chi(\complement E_m)$. Because of (26), we have $\|f\|_\gamma \leq p^{r+1}$. Condition (b) for $s = m$ means that the interval $\Delta = [v + 1, v + p^r]$ is normal. Thus, we can apply the lemma and define the numbers $l = l_{m+1}, k = k_{m+1}$ so that conditions (i) and (ii) of the lemma are satisfied. Taking into account (26) and (27) we observe that these will satisfy conditions (a)–(c) when $s = m + 1$. Therefore the inductive step is completed.

Because of the strict inclusion of the intervals δ_s, the process stops after a finite number of steps. Thus, we have defined a number $q \geq 1$ and the sequences $\{l_s\}, \{k_s\}$, the functions $\{F_s\}, \{f_s\}$, the sets $\{E_s\}, \{U_s\}$ and the numbers $\{I_s\}, \{J_s\} \ (s \leq q)$ so that conditions (a)–(d) are satisfied.

From (26) it immediately follows that

$$f_s(t) = F_s(t) \quad (t \in \complement E_{s-1}) \quad (1 \leq s \leq q) . \tag{28}$$

Observe the relations between the sets:

$$E_{s-1} \subset E_s; \quad E_s \setminus E_{s-1} = U_s \cap \complement E_{s-1} . \tag{29}$$

The first of these results from the inequality

$$|F_s(t)| \geq |F_{s-1}(t)| - \left| \sum_{i=l_{s-1}+1}^{l_s} c_i \phi_i(t) \right| > p^{k_s-i+1} - p^{k_s-i} > p^{k_s+1} \quad (t \in E_{s-1}),$$

where we have used the relations (26), (1), (3) and (b), in that order. The second follows from (28) and (26).

We shall now prove by induction the following inequality:

$$I_m \geq \sum_{1 \leq s < m} J_s \left(1 - \frac{2}{p} - \cdots - \frac{2}{p^{m-s}} \right) + J_m \quad (1 \leq m \leq q). \tag{30}$$

When $m=1$ it is obvious since $f_1 = F_1, U_1 = E_1$; that is, $J_1 = I_1$. Suppose inequality (30) is true for some value $m < q$. Then we have, taking into account (26)–(29),

$$
\begin{aligned}
I_{m+1} &= \int_{E_{m+1}} |F_{m+1}| d\mu = \int_{E_m} |F_{m+1}| d\mu + \int_{E_{m+1} \setminus E_m} |F_{m+1}| d\mu \\
&\geq I_m - \int_{E_m} \left| \sum_{i=l_m+1}^{l_{m+1}} c_i \phi_i \right| d\mu + \int_{E_{m+1} \setminus E_m} |f_{m+1}| d\mu \\
&\geq I_m - \int_{E_m} \left| \sum_{i=l_m+1}^{l_{m+1}} c_i \phi_i \right| d\mu + \int_{U_{m+1}} |f_{m+1}| d\mu - \int_{E_m} |f_{m+1}| d\mu \\
&= I_m + J_{m+1} - 2 \int_{E_m} \left| \sum_{i=l_m+1}^{l_{m+1}} c_i \phi_i \right| d\mu .
\end{aligned}
\tag{31}
$$

Further, when $t \in E_s, s \leq m$, we have, because of (a) and (26),

$$\left| \sum_{i=l_m+1}^{l_{m+1}} c_i \phi_i(t) \right| \leq p^{k_m} = \frac{1}{p^{k_s-k_m+1}} p^{k_s+1}$$

$$< \frac{1}{p^{k_s-k_m+1}} |F_s(t)| \leq \frac{1}{p^{m-s+1}} |F_s(t)|.$$

Therefore because of (29), (26), (27),

$$\int_{E_m} \left| \sum_{i=l_m+1}^{l_{m+1}} c_i \phi_i \right| d\mu = \sum_{s=1}^{m} \int_{E_s \setminus E_{s-1}} \left| \sum_{i=l_m+1}^{l_{m+1}} c_i \phi_i \right| d\mu \leq \sum_{s=1}^{m} \frac{1}{p^{m-s+1}} \int_{E_s \setminus E_{s-1}} |F_s| d\mu,$$

$$\sum_{s=1}^{m} \frac{1}{p^{m-s+1}} \int_{E_s \setminus E_{s-1}} |f_s| d\mu \leq \sum_{s=1}^{m} \frac{1}{p^{m-s+1}} J_s.$$

Hence because of (31) and the inductive assumption we obtain

$$I_{m+1} \geq \sum_{1 \leq s < m} J_s \left(1 - \frac{2}{p} - \cdots - \frac{2}{p^{m-s}} - \frac{2}{p^{m-s+1}} \right) + J_m \left(1 - \frac{2}{p} \right) + J_{m+1}.$$

Therefore inequality (30) is true for all $1 \leq m \leq q$. Letting $m=q$, and taking (c) into consideration, we obtain

$$\int_{E_q} |F_q| d\mu \geq \frac{1}{2} \sum_{s=1}^{q} J_s \geq \frac{1}{2\lambda} \sum_{s=1}^{q} [\gamma(\delta_{s-1}) - \gamma(\delta_s)] = \frac{1}{2\lambda} [\gamma(\delta_0) - \gamma(\delta_q)].$$

In the case $\gamma(\delta_q) < \frac{1}{2}\gamma(\delta_0)$ we take into account (24) and (25), and get

$$\int_X \left| \sum_{i=l_0+1}^{l_q} c_i \phi_i(t) \right| d\mu > \frac{1}{4\lambda}\gamma(\delta_0) \geq \frac{1}{8\lambda n}(r_0+1)\sum_{k=1}^{n}|c_k|^2 > \frac{1}{8\lambda}\frac{\log_p n}{n}\sum_{k=1}^{n}|c_k|^2.$$

In the case $\gamma(\delta_q) \geq \frac{1}{2}\gamma(\delta_0)$, using condition (d), we have

$$\int_X \left| \sum_{i=l_q}^{l_q+p^{k_q}} c_i \phi_i \right| d\mu \geq \frac{1}{\left\| \sum\limits_{i=l_q}^{l_q+p^{k_q}} c_i \phi_i \right\|_\infty} \sum_{i\in\delta_q}|c_i|^2 \geq \frac{\log_p n}{p^{3p}}\cdot\frac{1}{n}\sum_{k=1}^{n}|c_k|^2. \tag{32}$$

In both cases it is clear that we get inequality (2) with the constant $K = p^{-5p}$.

This completes the proof of Theorem 1.

Notice the following direct consequences.

1. The Littlewood hypothesis. This hypothesis (see [51]) essentially concerns the behavior of the Lebesgue constants of an arbitrarily ordered trigonometric system. It can be formulated in the following way: for any rearrangement $\{v_k\}$ of the natural numbers the relation

$$\liminf_{n\to\infty}\frac{I_n}{\ln n} > 0, \quad \text{where} \quad I_n = \int_{-\pi}^{\pi}\left| \sum_{k=1}^{n} e^{iv_k t} \right| dt,$$

is satisfied. This hypothesis was the subject of the investigations of a series of authors. In particular, Salem [116] showed that with special assumptions on the sequence $\{v_k\}$ the following relation holds:

$$\limsup_{n\to\infty}\frac{I_n}{\sqrt{\ln n}} > 0.$$

Important progress was made by P. Cohen [23]: for any rearrangement $\{v_k\}$

$$\liminf_{n\to\infty}\frac{I_n}{(\ln n/\ln\ln n)^{1/8}} > 0. \tag{33}$$

The result of Cohen, in particular, for the first time established that the trigonometric system does not form a basis in the space C for any ordering. By adding somewhat to the method Cohen used, Davenport [24] was able to replace the exponent $1/8$ in (33) by $1/4$.

From our inequality (2) it follows at once that *any bounded orthonormal system satisfies the relation*

$$\limsup_{n\to\infty}\frac{1}{\ln n}\int_X\left| \sum_{k=1}^{n}\phi_k(t) \right| d\mu > 0.$$

Thus the Littlewood hypothesis turns out to be true for any bounded orthonormal system, but with an upper limit substituted for the lower limit. In the class of systems mentioned, such a substitution is necessary, as is shown by the example of the Walsh system $\{w_i\}$. This system satisfies the equality

$$\int_0^1 \left| \sum_{i=1}^{2^k} w_i(t) \right| dt = 1 \qquad (k = 1, 2, \ldots)$$

(it is sufficient to notice that the left hand side equals $L_{2^k}^w(0) = L_{2^k}^\chi(0)$, see Lemma 1 § 2).

2. It follows from Theorem 1 that *if a series* $\sum c_n \phi_n$, *where* $\{\phi_n\}$ *is an ONS and* $\sup_n \|\phi_n\|_\chi < \infty$, *has partial sums bounded in the metric L, then its coefficients converge in mean to zero:* $\sum_{k=1}^n |c_k|^2 = o(n)$. One might conjecture that this takes place not only in mean but $c_k = o(1)$. For the trigonometric system this is precisely the case (Helson, see [10]). However, the proof depends greatly on this particular system and uses its relationship to the theory of boundary values of analytic functions. It turns out that there is no similar result for rearrangements of the trigonometric system. The following proposition is true (Sidon [121]): *there exists a series* $\sum \alpha_k \cos kt$, $\limsup |\alpha_k| > 0$, *which after some rearrangement of its terms is bounded in* $L[0, \pi]$.

For the proof we look at the Riesz product $R_n = \prod_{k=1}^n (1 + 1/3 \cos 3^k t)$. It is easy to see (see [171]) that R_n, for each n, is a trigonometric polynomial of the form

$$R_n = \sum_{l=0}^{2 \cdot 3^n} \alpha_l \cos lt = R_{n-1} + \sum_{l = 2 \cdot 3^{n-1} + 1}^{2 \cdot 3^n} \alpha_l \cos lt, \qquad |\alpha_l| \le 1.$$

Define by induction a rearrangement of the terms of this polynomial. Let the polynomial R_{n-1} be already ordered in the form

$$R_{n-1} = \sum_{l=0}^{2 \cdot 3^{n-1}} \alpha_{v_l} \cos v_l t.$$

Then the equation

$$R_n = R_{n-1} + \tfrac{1}{3} R_{n-1} \cos 3^n t = \sum_{l=0}^{2 \cdot 3^{n-1}} \alpha_{v_l} \cos v_l t$$

$$+ \tfrac{1}{3} \sum_{l=0}^{2 \cdot 3^{n-1}+1} \alpha_{v_l} \tfrac{1}{2} [\cos(3^n + v_l)t + \cos(3^n - v_l)t] \equiv \sum_{l=0}^{2 \cdot 3^n} \alpha_{v_l} \cos v_l t \qquad (34)$$

defines an ordering of the terms of R_n. Look at the series

$$\sum_{l=0}^r \alpha_{v_l} \cos v_l t$$

with partial sums S_l. Clearly, $S_{2 \cdot 3^n}(x) = R_n(x) > 0$. Defining $\lambda_n = \max\limits_{0 \le m \le 2 \cdot 3^n} \int\limits_0^\pi |S_m|$. it is easy to conclude from (34) the inequality

$$\lambda_n \le \max \left\{ \lambda_{n-1} \cdot \int\limits_0^\pi |R_{n-1}| dt + \tfrac{1}{3} \lambda_{n-1} + \frac{\pi}{6} \right\} \le \max \left\{ \lambda_{n-1}, 4 + \frac{\lambda_{n-1}}{3} \right\}.$$

Hence it follows that $\sup\limits_n \lambda_n \le 6$. At the same time the coefficients of $\cos 3^n t$ are equal to $1/3$.

§ 2. The Logarithmic Growth of the Lebesgue Functions. Divergence of Fourier Series

The inequality of the preceding section permits us to reach the conclusion that if an orthonormal system is bounded then its Lebesgue functions are not bounded and the order of magnitude of their growth is no less than logarithmic.

Theorem 1. *Let* $\{\phi_n\}$ *be an ONS on* $X, \mu X < \infty$,

$$|\phi_n(x)| \le M \qquad (n = 1, 2, \ldots; x \in X). \tag{1}$$

Then the Lebesgue functions L_n *satisfy the condition*

$$\limsup_{n \to \infty} \frac{1}{\ln n} L_n(x) > 0 \qquad \left(x \in E, \mu E \ge \frac{1}{M^2} \right). \tag{2}$$

We define

$$E = \left\{ x \in X : \limsup_{n \to \infty} \frac{1}{n} \sum_{k=1}^n |\phi_k(x)|^2 > 0 \right\}. \tag{3}$$

Then

$$\frac{1}{n} \sum_1^n |\phi_k(x)|^2 = o(1), \qquad x \in \complement E.$$

Hence after integrating and using assumption (1), we get

$$1 - M^2 \mu E \le \frac{1}{n} \sum_{k=1}^n \left[1 - \int\limits_E |\phi_k|^2 dt \right] = \frac{1}{n} \int\limits_{\complement E} \sum_{k=1}^n |\phi_k|^2 dt = o(1).$$

That is, $\mu E \ge \dfrac{1}{M^2}$.

Fixing $x \in E$ we have, on the basis of the theorem of § 1,

$$\max_{1 \le k \le n} |\phi_k(x)| \max_{1 \le k \le n} L_k(x) \ge \frac{K}{M} \frac{\sum\limits_{k=1}^n |\phi_k(x)|^2}{n} \ln n.$$

Hence because of (1) and (3), it follows that (2) is true.

A basic consequence of Theorem 1 is the following.

Theorem 2 [97]. *No uniformly bounded ONS can form a basis in the space* $C[a,b]$.

What is more, the following generalization of the theorem of du Bois-Reymond is true [97].

Theorem 3. *If the ONS* $\{\phi_n\}$ *is uniformly bounded, then there exists a continuous function whose Fourier series diverges at some point.*

This follows from the relation

$$L_n(x) \neq O(1) \qquad (x \in E), \tag{4}$$

which is a result of (2). Using (2) to the complete extent, it is possible to strengthen the result by estimating the rate of divergence of the Fourier series.

Theorem 4. *Let* $\{\phi_n\}$ *be an ONS satisfying condition* (1). *Then for any sequence* $\omega(n) = o(\ln n)$ *there exists a function* $f \in C$ *such that at some point* x *(and even on a set of the power of the continuum) the relation*

$$\limsup_{n \to \infty} \frac{|S_n(f; x)|}{\omega(n)} = \infty \tag{5}$$

is satisfied.

Here $S_n(f; x) = \sum_{k=1}^{n} c_k(f) \phi_k(x)$. We fix $x \in E$ and look at the linear functionals

$$\Phi_n = \frac{1}{\omega(n)} S_n(f; x).$$

Clearly, $\|\Phi_n\| = \frac{1}{\omega(n)} L_n(x)$, and so, because of (2), $\|\Phi_n\| \neq O(1)$. By the Banach-Steinhaus theorem the set $A_x \subset C$ of functions for which (5) is true at the point x is of the second category in C (in fact, its complement is a countable union of nowhere dense sets). Because of the C-property of Lusin we can choose a perfect set $E' \subset E, \mu E' > 0$, on which every ϕ_k is continuous. Let the set $\{x_m\}$ be dense in E'. Clearly, the set $A = \bigcap_m A_{x_m}$ is not empty (it is of the second category in C). Let $f \in A$ and let U_f be the set of all points satisfying (5). It is easy to see that it is of the form G_δ and is also dense in E' ($x_m \in U_f$). Therefore U_f is uncountable. We now use a theorem of P.S. Alexandrov: every uncountable Borel set in a Euclidean space has the power of the continuum.

Thus, for the majority (in the category sense) of functions in the space C, the Fourier series with respect to an arbitrary bounded orthonormal system diverges on a maximally large (in the sense of power) set of points.

Nothing can be asserted about the measure of the set of points of divergence, since according to the theorem of Carleson for the trigonometric system, convergence takes place almost everywhere for any $f \in C$. On the other hand, for the trigonometric system rearranged in a certain manner, the Fourier series of some continuous function diverges almost everywhere [93] (see Chap. III).

From the case of the trigonometric system we see that the estimate of the rate of divergence contained in (5) is the sharpest possible. Indeed, for this system we have the relation (see [10])

$$\|S_n(f)\| = o(\ln n) \qquad (\forall f \in C).$$

In connection with Theorems 1–3 the following questions, concerning the possibility of strengthening these results in various directions, arise quite naturally.

1. Is it possible in relation (2) to replace the limit superior by a limit inferior; that is, is it possible to give a lower bound for the Lebesgue functions not merely for infinitely many, but for all numbers n, as can be done for the trigonometric system?

2. Is it possible to assert that (2) (or the even weaker condition (4)) holds almost everywhere (assuming, of course, that the system is complete)? In other words, is it true that the divergence of a Fourier series with respect to a complete bounded system can take place at almost every point of the interval $[a,b]$?

3. Is there a certain smoothness condition such that for any complete bounded system there exists a function having this smoothness and still satisfying the conclusion of Theorem 3?

These questions will be examined below. The results obtained here describe the degree of finality of the basic theorems and, to a certain extent, outline the boundaries of the possibility of extending to general bounded systems the local properties of trigonometric Fourier series.

Subsequences of convergence. In reference to the first question, a negative answer is given by the classical Walsh system. The subspaces generated by the Haar functions $\{\chi_i\}$ and by the Walsh functions $\{w_i\}$, where $1 \le i \le 2^k$, coincide. Therefore, considering the uniform boundedness of the Lebesgue functions for the Haar system, we conclude that for the Walsh system the Lebesgue functions with indices $v_k = 2^k$ satisfy the condition

$$L_{v_k}(x) = O(1) \qquad \text{(uniformly in } x). \tag{6}$$

This means that $v_k = 2^k$ is a subsequence of convergence for this system; that is, for any function $f \in C$ the Fourier partial sums $S_{v_k}(f)$ converge uniformly to f.

It turns out, however, that for a bounded system relation (6) can be satisfied only for sparse sequences of indices; roughly speaking, only for sequences lacunary in the sense of Hadamard, that is, growing like a geometric progression. More precisely, the following proposition holds.

Theorem 5. *If* $\{\phi_i\}$ *is an orthonormal system satisfying condition* (1), *and if* $\{v_k\}$ *is a sequence of indices satisfying the condition*

$$\lim_{k \to \infty} \frac{v_{k+1}}{v_k} = 1, \tag{7}$$

then

$$L_{v_k}(x) \ne O(1) \qquad (\forall x \in E, \mu E > 0). \tag{8}$$

Thus, for some $f \in C$ the partial sums S_{v_k} diverge at the point x. In particular, this takes place for $v_k = k^s$ or $v_k = [2^{k^a}]$, $\alpha < 1$ (but not for $\alpha = 1$).

Proof. To each natural number $l > 1$ is associated a number N_l such that $v_{k+1} < \left(1 + \dfrac{1}{2l}\right) v_k$ $(v_k > N_l)$. It is evident that then each of the half-closed intervals $\left(N_l\left(1 + \dfrac{i-1}{l}\right), N_l\left(1 + \dfrac{i}{l}\right)\right]$ $(1 \le i \le l)$ contains at least one element $v_i^{(l)}$ of the sequence $\{v_k\}$. Clearly,

$$\int_x \frac{1}{v_l^{(l)} - v_1^{(l)}} \sum_{v = v_1^{(l)} + 1}^{v_l^{(l)}} \phi_v^2(x) \, dx = 1; \quad \left\| \sum_{v = v_1^{(l)} + 1}^{v_l^{(l)}} \phi_v^2(x) \right\|_x \le M^2(v_l^{(l)} - v_1^{(l)}).$$

Therefore

$$\sum_{v = v_1^{(l)} + 1}^{v_l^{(l)}} \phi_v^2(x) \ge \frac{1}{2\mu X}(v_l^{(l)} - v_1^{(l)}) \quad \left(x \in E_l, \mu E_l > \frac{1}{2M^2}\right).$$

Let $E = \limsup\limits_{l \to \infty} E_l$. Clearly, $\mu E > 0$. For each $x \in E_l$ we have

$$\sum_{v = v_1^{(l)} + 1}^{v_l^{(l)}} \phi_v(x)\phi_v(t) = \sum_{i=2}^{l} \sum_{v = v_{i-1}^{(l)} + 1}^{v_i^{(l)}} \phi_v(x)\phi_v(t) = \sum_{i=2}^{l} d_i(x)\Phi_i^{(x)}(t),$$

where

$$d_i(x) = \left[\sum_{v = v_{i-1}^{(l)} + 1}^{v_i^{(l)}} \phi_v^2(x)\right]^{1/2}; \quad \Phi_i^{(x)}(t) = \frac{1}{d_i(x)} \sum_{v = v_{i-1}^{(l)} + 1}^{v_i^{(l)}} \phi_v(x)\phi_v(t)$$

and the functions $\{\Phi_i^{(x)}\}$ are orthonormal.
Clearly,

$$\|\Phi_i^{(x)}\|_x \le M(v_i^{(l)} - v_{i-1}^{(l)})^{1/2}; \quad |d_i(x)| \le M(v_i^{(l)} - v_{i-1}^{(l)})^{1/2}.$$

Therefore the fundamental inequality (§ 1) gives

$$\max_{2 \le m \le l} \left\| \sum_{i=2}^{m} d_i \Phi_i^{(x)} \right\|_1 \ge \frac{K}{M^2 \max\limits_i (v_i^{(l)} - v_{i-1}^{(l)})} \frac{1}{l-1} \sum_{i=2}^{l} d_i^2(x) \ln(l-1)$$

$$\ge \frac{K}{M^2 \dfrac{N_l}{l}} \sum_{v = v_1^{(l)} + 1}^{v_l^{(l)}} \phi_v^2(x) \frac{1}{l-1} \ln(l-1)$$

$$\ge \frac{K}{4 M^2 \mu X} \ln(l-1).$$

Thus, for each $x \in E$ and for infinitely many numbers l we have

$$\max_{\substack{N1 < \mu < \mu' \le 2 N_l \\ \mu, \mu' \in \{v_k\}}} \int_x \left| \sum_{v = \mu + 1}^{\mu'} \phi_v(x)\phi_v(t) \right| dt \ge \frac{K}{4 M^2 \mu X} \ln(l-1).$$

Therefore (8) is true.

This result shows that if a bounded orthonormal system generates a so-called basis of subspaces (see [41]) in the space C, then the dimension of these subspaces grows sufficiently rapidly.

The set of points of growth of the Lebesgue functions. Passing on to the second question posed above, we notice that for any ordering of the trigonometric system (or the Walsh system), the set E in relation (3) has full measure; that is, (2) is satisfied almost everywhere (even everywhere). It turns out that it is possible to construct a complete bounded ONS ϕ, having better properties: the growth of its Lebesgue functions is localized in a small neighborhood. Thus, outside this neighborhood, Fourier series converge.

Theorem 6 [97]. *For any* δ, $0<\delta<1$ *there exists a uniformly bounded ONS* $\{\phi_n\}$, *closed in* $C[0,1]$, *satisfying*

$$L_n^\phi(x)<K_1 \qquad (\delta\leq x\leq 1).$$

Below, the K_i will be constants depending only on δ.

Lemma 1. *If two ONS* $\{\alpha_i\}$ *and* $\{\beta_i\}$ $(1\leq i\leq n)$ *are equivalent, then* $\mathcal{D}_n^\alpha(s,t)\equiv\mathcal{D}_n^\beta(s,t)$ $(\mathcal{D}_n^\psi$ *is the Dirichlet kernel of the system* ψ*).*

This fact can be verified directly.

Lemma 2. *Let* $\{\psi_j\}$ $(0\leq j\leq n)$ *be an ONS,* $\|\psi_0\|_\infty\leq\sqrt{n}$, $\|\psi_j\|_\infty<\lambda$ *for* $j>0$. *Then one can pass to an equivalent ONS* $\{\phi_j\}$, $\|\phi_j\|_\infty\leq C(\lambda)$ $(0\leq j\leq n)$ *by means of some matrices depending only on* n.

This result is simple too (for example, one can use the matrices A of Chap. IV § 1).

Let $\psi^{(0)}=\{\psi_k^{(0)}\}$ $(k\geq 1)$ be the system orthonormal on $[\delta,1]$ resulting from the Haar system by a linear substitution of variables and normalization. Let $\psi=\{\psi_k\}$ be the ONS on $[0,\delta)$ resulting from the system $\{\cos(k-1)t\}$ $(t\in[0,\pi))$ by an analogous procedure. Outside the specified intervals the functions are assumed to be equal to zero. The following properties are clear:
  (i)   the system $\psi\cup\psi^{(0)}$ is orthonormal on $[0,1]$ and closed in $C[0,1]$;
  (ii) $\|\psi_k^{(0)}\|_\infty<K_2 k^{1/2}$; $\|\psi_k^{(0)}\|_1<K_2 k^{-1/2}$;
  (iii) $\|\psi_k\|_\infty<K_3$;

  (iv) $\int\limits_0^1\left|\sum\limits_{s=1}^k\psi_s^{(0)}(x)\psi_s^{(0)}(t)\right|dt=1$;

  (v) $\int\limits_0^1\left|\sum\limits_{i=m+1}^{m+p}\psi_i\right|dt<K_4\ln(p+1)$ $(\forall m,p)$.

Divide the system ψ into blocks $\{\psi_k^{(j)}\}$ $(1\leq j\leq k^2+1; k=1,2,\dots)$ and let

$$\phi_k^{(i)} = \sum_{j=0}^{k^2} b_{ij}^{(k^2)}\psi_k^{(j)} \tag{8'}$$

where the transition matrices have the form

$$
B_n = \|b_{ij}^{(n)}\| =
\begin{pmatrix}
0 & \dfrac{1}{\sqrt{n}} & \cdot & \cdot & \cdot & \cdot & \cdot & \dfrac{1}{\sqrt{n}} \\[6pt]
\dfrac{1}{\sqrt{n}} & 1-\dfrac{1}{n} & & & & & & -\dfrac{1}{n} \\[6pt]
\cdot & & \ddots & & & & & \\[6pt]
\cdot & & & \ddots & & & & \\[6pt]
\cdot & & & & \ddots & & & \\[6pt]
\dfrac{1}{\sqrt{n}} & -\dfrac{1}{n} & & & & 1-\dfrac{1}{n}
\end{pmatrix}.
\tag{9}
$$

These matrices are orthogonal. Thus the system $\{\phi_k^{(i)}\}$ $(0\le i\le k^2; k=1,2,\ldots)$ is orthonormal on $[0,1]$ and is equivalent to the system $\{\psi_k^{(j)}\}$ $(0\le j\le k^2; k=1,2,\ldots)$. On the basis of Lemma 2 we can choose an ONS $\{g_k\}$, $\|g_k\|_\infty < M$, equivalent to the system $\{\phi_k^{(0)}, \psi_k^{(k^2+1)}\}$. Define $\phi_k^{(k^2+1)} = g_k$. Then the system $\phi = \{\phi_k^{(i)}\}$ $(1\le i\le k^2+1; k=1,2,\ldots)$ is orthonormal on $[0,1]$ and is equivalent to $\psi \cup \psi^{(0)}$, and consequently, because of (i), is closed in C. Furthermore, from (ii) and (iii) we have

$$
\sup_{x\in[0,1]} |\phi_k^{(i)}(x)| \le \frac{1}{k}\|\psi_k^{(0)}\|_\infty + 2K_3 = O(1) \qquad (1\le i\le k^2).
$$

Therefore ϕ is uniformly bounded.

Notice the following relations resulting from (ii), (v), (8′), and (9):

$$
\int_0^1 \left| \sum_{i=1}^l \phi_k^{(i)} \right| dt = \frac{l}{k}\int_\delta^1 |\psi_k^{(0)}| dt + \int_0^\delta \left| \sum_{i=1}^l \left[\psi_k^{(i)} - \frac{1}{k^2}\sum_{j=1}^{k^2} \psi_k^{(j)} \right] \right| dt
$$

$$
\le K_2 k^{1/2} + 4K_4 \ln(k+1) < K_5 k^{1/2} \qquad (1\le i\le k^2);
$$

$$
\sum_{i=1}^{s^2+1} \phi_s^{(i)}(x)\phi_s^{(i)}(t) = \sum_{i=0}^{s^2} \phi_s^{(i)}(x)\phi_s^{(i)}(t) = \sum_{j=0}^{s^2} \psi_s^{(i)}(x)\psi_s^{(i)}(t) = \psi_s^{(0)}(x)\psi_s^{(0)}(t) \qquad (x\ge\delta)
$$

(here it has been taken into account that $\phi_s^{(0)}(x) = \phi_s^{(s^2+1)}(x) = 0$, and Lemma 1 has been used). Therefore, after considering (iv), we have the following, for any $k,l\in[1,k^2]$ and $x\in[\delta,1]$:

$$
\int_0^1 \left| \sum_{s=1}^k \sum_{i=1}^{s^2+1} \phi_s^{(i)}(x)\phi_s^{(i)}(t) + \sum_{i=1}^l \phi_k^{(i)}(x)\phi_k^{(i)}(t) \right| dt
$$

$$
\le \int_0^1 \left| \sum_{s=1}^k \psi_s^{(0)}(x)\psi_s^{(0)}(t) \right| dt + \frac{|\psi_k^{(0)}(x)|}{k}\int_0^1 \left| \sum_{i=1}^l \phi_k^{(i)}(t) \right| dt
$$

$$
\le 1 + \frac{k^{1/2}}{k}\cdot K_5\cdot k^{1/2} = K_1.
$$

Therefore the Lebesgue functions of the system ϕ are uniformly bounded on the interval $[\delta,1]$ and the proof of Theorem 6 is complete.

It is easy to see from the proof that

$$M(\delta) \equiv \sup_n \|\phi_n\|_x = O\left(\frac{1}{\sqrt{\delta}}\right), \qquad \delta \to 0.$$

This shows that the estimate (2) in terms of M of the measure of the set of points of growth of the Lebesgue functions is in the sense of order of magnitude exact in the class of complete bounded systems.

As has already been noted, the following assertion is true for the system ϕ just constructed: $\forall f \in C[0,1]$ *the Fourier series* $\sum c_n \phi_n$ *converges to* f *uniformly on* $[\delta,1]$. Here it is even sufficient to assume that f is summable on $[0,1]$ and continuous for all $x \geq \delta$ (see [97]).

The smoothness of functions with divergent Fourier series. According to the theorem of Dini and Lipschitz, if a function f has a modulus of continuity of the order $o\left(\dfrac{1}{|\ln \delta|}\right)$, then it can be expanded as a uniformly convergent trigonometric Fourier series. This result cannot be improved: there exists a function f_0, $\omega(\delta, f_0) = O\left(\dfrac{1}{|\ln \delta|}\right)$, with Fourier series diverging at some point (see [171]). The existence of such an example is closely connected with the logarithmic growth of the Lebesgue constants. Therefore, keeping in mind Theorem 1, it is possible to anticipate that a similar situation occurs even in the general case. It turns out that this is actually the case, but only with a few additional conditions. It is also important that the functions ϕ_n do not oscillate too rapidly. Without attempting complete generality, we illustrate this by the following theorem.

Theorem 7. *Let* ϕ *be an ONS satisfying condition* (1) *and*

$$\|\phi_n\|_V = O(n^s) \tag{10}$$

for some s. *Then there exists a function* $f \in H^{\omega_0}, \omega_0 = |\ln \delta|^{-1}$, *whose Fourier series diverges at some point.*

Fix $x_0 \in E$ from (3) and let $d_k(t) = \sum_{i=n_k+1}^{n_k+v_k} \phi_i(x_0)\phi_i(t)$, where the indices n_k, v_k are chosen by induction. At each step the index v_k,

$$v_k > \max(n_k, 2^{k^2}), \tag{11}$$

is chosen, on the basis of (2), so that

$$\|d_k\|_1 > c \ln v_k, \qquad c = c(x_0) > 0. \tag{12}$$

From (1) and (10) it follows that $\|d_k\|_V < K v_k^{s+1}$, $\|d_k\|_x < K v_k$. As is easy to see, this allows us to define a step function σ_k with intervals of constancy $\{\rho_i^{(k)}\}$ $(1 \leq i \leq i_k < K v_k^{s+1})$ so that σ_k satisfies the conditions

$$\|\sigma_k\|_x < K v_k, \qquad \|\sigma_k - d_k\|_x < 1. \tag{13}$$

For each interval $\rho \subset X \equiv [a,b]$, denote by $\Theta_\rho(x)$ the function equal to zero outside this interval, one at the center of the interval, and linear on each half-interval. Set

$$f_k(x) = \frac{\operatorname{sign} \sigma_k(x)}{\ln v_k} \sum_{i \in J_k} \Theta_{\rho_i^{(k)}}(x), \tag{14}$$

where $J_k = \left\{ i : |\rho_i^{(k)}| > \frac{1}{v_k^{s+2}} \right\}$. It is easy to see that

$$\omega(f_k; \delta) \leq \omega_k(\delta) \equiv \begin{cases} \dfrac{2 v_k^{s+2}}{\ln v_k}\, \delta, & \delta \leq v_k^{-(s+2)}, \\[2ex] \dfrac{2}{\ln v_k}, & \delta > v_k^{-(s+2)}. \end{cases} \tag{15}$$

It is not difficult to verify that if v_k grows sufficiently rapidly, that is,

$$\frac{v_{k+1}^{s+2}}{\ln v_{k+1}} > 2 \frac{v_k^{s+2}}{\ln v_k}, \tag{16}$$

then

$$\sum_1^\infty \omega_k(\delta) = O(\omega_0(\delta)). \tag{17}$$

Finally, the index n_{k+1} is chosen so as to satisfy the conditions

$$n_{k+1} > 3^{v_k}, \tag{18}$$

$$\sup_{n_{k+1} < n < m} \left| \sum_{i=n}^m c_i(F_k) \phi_i(x_0) \right| < \frac{1}{k}, \quad F_k = \sum_{j=1}^k f_j. \tag{19}$$

In reference to the last condition, we notice that if for some k it is not possible to choose n_{k+1} such that the condition is satisfied, then the function $f = f_j$ for some $j \leq k$ satisfies the requirements of the theorem, because of (15).

Suppose $f = \sum_1^\infty f_k$ (the series converges in C since $\|f_k\|_C \leq \dfrac{1}{\ln v_k} < \dfrac{1}{k^2}$ because of (14) and (11)). Conditions (11) and (18) provide justification for (16) and therefore because of (15) and (17) we have $f \in H^{\omega_0}$. Further, for each k we have

$$\left| \sum_{i=n_k+1}^{n_k+v_k} c_i(f) \phi_i(x_0) \right| \geq \left| \sum_{i=n_k+1}^{n_k+v_k} c_i(f_k) \phi_i(x_0) \right| - \left| \sum_{i=n_k+1}^{n_k+v_k} c_i(F_{k-1}) \phi_i(x_0) \right|$$

$$- \left| \sum_{i=n_k+1}^{n_k+v_k} c_i \left(\sum_{l=k+1}^\infty f_l \right) \phi_i(x_0) \right|.$$

The second sum on the right is estimated by (19). The modulus of the last sum is equal to

$$\left| \int_X \sum_{l=k+1}^\infty f_l(t) d_k(t)\, dt \right| \leq \|d_k\|_1 \sum_{k+1}^\infty \|f_l\|_\infty \leq \left(\sum_{i=n_k+1}^{n_k+v_k} \phi_i^2(x_0) \right)^{1/2} (\mu X)^{1/2} \sum_{l=k+1}^\infty \frac{1}{\ln v_l};$$

that is, $\leq O(\sqrt{v_k})\dfrac{1}{\ln v_{k+1}} = o(1)$, because of (1), (18), and (11). Finally, as a consequence of (13) and (14), we have

$$\left| \sum_{i=n_k+1}^{n_k+v_k} c_i(f_k)\phi_i(x_0) \right| = \left| \int_X f_k d_k dt \right| \geq \left| \int_X f_k \sigma_k dt \right| - \frac{\mu X}{\ln v_k}$$

$$= \frac{1}{\ln v_k} \sum_{i \in J_k} \int_{\rho_i^{(k)}} \Theta_{\rho_i^{(k)}} |\sigma_k| dt - \frac{\mu X}{\ln v_k}$$

$$= \frac{1}{2\ln v_k}\left[\sum_{i \in J_k} \int_{\rho_i^{(k)}} |\sigma_k| dt - 2\mu X \right] \geq \frac{1}{2\ln v_k}\left[\|\sigma_k\|_1 - \sum_{i \notin J_k} \int_{\rho_i^{(k)}} |\sigma_k| dt - 2\mu X \right]$$

$$\geq \frac{1}{2\ln v_k}\left[\|d_k\|_1 - 4\mu X - \frac{i_k}{v_k^{s+2}} \|\sigma_k\|_\infty \right] \geq \frac{1}{2\ln v_k}\left[c\ln v_k - O(1) \right]$$

(the last is from (12), (13) and the estimate for i_k). Finally, we obtain

$$\left| \sum_{i=n_k+1}^{n_k+v_k} c_i(f)\phi_i(x_0) \right| > \frac{c}{2} - o(1),$$

and the proof of the theorem is complete.

Notice that the unbounded divergence of the Fourier series cannot be asserted
here, since we have already in the trigonometric case, for $f \in H^{\omega_0}$,

$$\|S_n(f)\| \leq \|S_n(f - P_n)\| + \|P_n\| \leq K \ln n\, E_n(f) + \|P_n\|$$

$$= O\left(\ln n\, \omega_0\left(\frac{1}{n} \right) \right) + O(1) = O(1)$$

(here $P_n(f)$ is the trigonometric polynomial of order n best approximating f,
$E_n(f) = \|f - P_n\|$; the theorem of Jackson was used).

Theorem 7 shows that, for a bounded order of oscillation of the functions ϕ_n
forming the bounded system, we have, in the question of the expansion of the
function f as a uniformly convergent Fourier series, this same maximal smoothness
that occurred also in the trigonometric case. If rapid oscillations are allowed then it
turns out that it is possible to construct systems giving uniform Fourier expansions
in any pre-assigned compact subset of the space C.

Theorem 8. *For any modulus of continuity $\omega(\delta)$ there exists a uniformly bounded
ONS $\{\phi_n\}$, with respect to which every function $f \in H^\omega$ expands as a uniformly
convergent Fourier series.*

The system ψ. We introduce an auxiliary system $\psi = \{\psi_n\}$ [98]. Its properties
are close to those of the Haar system, differing from it in one respect: it contains
an infinite uniformly bounded subsystem. Suppose

$$\psi_n = \chi_n \quad (n=1,2), \qquad \psi_n = \chi_l r_{k+1} \quad (n>2), \tag{20}$$

where the indices $l(n), k(n)$ are determined by the conditions $2^k < n \leq 2^{k+1}$,
$l = n - 2^k$ (define the functions ψ_n regularly at the points of discontinuity).

This system has the following properties.

(i) The system ψ is orthonormal on $[0.1]$.

(ii) Every continuous function expands as a uniformly convergent Fourier series in this system.

(iii) The Rademacher system $\{r_k\}$ is a subsystem of ψ.

Property (i) is verified directly. Further, each function ψ_n $(2^k < n \leq 2^{k+1})$ is a linear combination of the Haar functions $\chi_k^{(j)}$ $(1 \leq j \leq 2^k)$. Because the dimensions coincide, it is clear that the converse is also true. This shows that the system ψ is closed in C and that for this system the Lebesgue functions with the indices 2^k coincide (by Lemma 1). Furthermore, it is clear that

$$\sum_{i=2^k+1}^{n} \psi_i(x)\psi_i(t) = r_{k+1}(x)r_{k+1}(t) \sum_{s=1}^{l} \chi_s(x)\chi_s(t) .$$

Thus

$$L_n^{\psi}(x) \leq L_{2^k}^{\psi}(x) + L_l^{\chi}(x) \leq 2 .$$

Notice finally that $\psi_2 = r_1$, $\psi_{2^k+1} = r_{k+1}$ $(k = 1, 2, \ldots)$.

We turn directly to the proof of Theorem 8. Let $\omega(\delta)$ be a given modulus of continuity. Select a sequence of odd numbers $\{q_k \uparrow\}$ that satisfies the condition $\sum \omega\left(\dfrac{1}{2^{q_k}}\right) < \infty$. Define $\psi' = \{r_{q_k}\}$, $\psi'' = \{r_{q_k+1}\}$, $\psi^0 = \psi \setminus (\psi' \cup \psi'')$. Because of the obvious estimate $\left|\int_0^1 f r_q \, dt\right| \leq \omega\left(\dfrac{1}{2^q}; f\right)$ the following condition is satisfied:

(a) $$\sum_{i:\psi_i \in \psi^0} |c_i(f)| < \infty ; \qquad c_i(f) = (f, \psi_i), f \in H^{\omega}.$$

Reorder the system $\psi^0 \cup \psi'$ and divide it into blocks $G_k = \{g_k^{(i)}\}$ $(0 \leq i \leq n_k; k = 1, 2, \ldots)$ that satisfy the conditions

(b) $$g_k^{(0)} \in \psi^0. \qquad \|g_k^{(0)}\| \equiv \sup_{x \in [0.1]} |g_k^{(0)}(x)| = \sqrt{n_k}.$$

(c) $g_k^{(j)} \in \psi'$ $\quad (1 \leq j \leq n_k)$.

(d) The order of the elements in ψ^0 is preserved.

(d), (ii) and (a) ensure that the following statement is true.

(e) Each function $f \in H^{\omega}$ has a uniformly convergent Fourier series in the system ψ^0.

We apply to each block G_k the linear transformation of the matrix B_{n_k} (9); that is, set

$$\phi_k^{(i)} = \sum_{j=0}^{n_k} b_{ij}^{(n_k)} g_k^{(j)} \qquad (0 \leq i \leq n_k; k = 1, 2, \ldots).$$

Also

$$\|\phi_k^{(i)}\| \leq |b_{i0}| \|g_k^{(0)}\| + \sum_{j>0} |b_{ij}| < 3, \quad i > 0; \quad \|\phi_k^{(0)}\| = \sqrt{n_k}.$$

Then, with the help of finite-dimensional transformations (Lemma 2), we pass to the bounded ONS $\{\phi_k'\}$ which is equivalent to the system $\{\phi_k^{(0)}\} \cup \psi''$. As a result we obtain the complete ONS $\phi = \{\phi_k^{(i)}, 1 \leq i \leq n_k\} \cup \{\phi_k'\}$.

Choosing the numbers $\{q_k\}$ increasing sufficiently rapidly, we can achieve (based on the estimate for (f, r_q)) that for every $f \in H^\omega$ the "small part" $f' = \sum_{i:\psi_i \in \psi^0} c_i(f)\psi_i$ would expand as a uniformly convergent series in the system ϕ.

So, in virtue of (e), to complete the proof it is sufficient to obtain the inequality

$$\|\sigma_\nu(g_k^{(0)})\| \le K\|g_k^{(0)}\|,$$

where $\sigma_\nu(g_k^{(0)}) = \sum_{i=0} (g_k^{(0)}, \phi_k^{(i)})\phi_k^{(i)}$, and K is an absolute constant.

Using properties of the matrices B_n that come easily from their definition, and considering assertions (b) and (c), we arrive at the following estimates.

$$\|\sigma_\nu(g_k^{(0)})\| = \left\| \sum_{i=0}^\nu b_{i0}\,\phi_k^{(i)} \right\| = \left\| \sum_{i=1}^\nu b_{i0} \sum_{j=0}^{n_k} b_{ij}g_k^{(j)} \right\|$$

$$= \left\| \sum_{j=0}^{n_k} \left(\sum_{i=1}^\nu b_{i0}b_{ij} \right) g_k^{(j)} \right\| \le \|g_k^{(0)}\| + \frac{1}{\sqrt{n_k}} \sum_{i,j>0} |b_{ij}| < 3\|g_k^{(0)}\|.$$

By comparison with Theorem 7 we notice that in the case $\omega(\delta)=|\ln\delta|^{-1}$ this construction gives the growth of $\|\phi_n\|_V$ as a geometric progression (in this case condition (a) is fulfilled for $g_k=k$). It would be interesting to investigate as precisely as possible the connection between the order of decrease of $\omega(\delta)$ and the minimal rate of oscillation of the functions of the system ϕ in Theorem 8.

Fourier series in the space L. The fundamental results of this section extend to the space $L(X)$. As is generally known (see [27]), the integral operator

$$A: L \to L, \ Af = \int_X K(s,t)f(t)dt, \ K \in L^\times (X \times X),$$

has norm $\|A\| = \left\| \int_X |K(s,t)|ds \right\|_\gamma$.

If an ONS ϕ satisfies condition (1) then it follows from relation (2) that

$$\|L_n\|_\infty > \alpha \ln n \quad (\alpha > 0)$$

for infinitely many n. Therefore for the operators $S_n(f) = \sum_{k=1}^n c_k(f)\phi_k$ we have

$$\limsup_{n \to \infty} \frac{1}{\ln n} \|S_n\| > 0.$$

The basic consequence of this is the following.

Theorem 9. *There does not exist a bounded ONS that forms a basis in L.*

For comparison we notice that by a theorem of M. Riesz the trigonometric system forms a basis in the space L^p for any $p \in (1, \infty)$.

Analogues of Theorems 4, 7, 8 are also true. For example: *if condition (1) is satisfied and $\omega(n)=o(\ln n)$, then there exists $f \in L$ for which*

$$\|S_n(f)\|_1 \ne O(\omega(n)).$$

§ 3. Series with Decreasing Coefficients

This section contains applications of the fundamental inequality of §1 to the investigation of series with monotone coefficients

$$\sum c_n \phi_n, \qquad c_n \downarrow 0, \tag{1}$$

with respect to uniformly bounded orthonormal systems

$$\|\phi_n\|_\infty < M \qquad (n = 1, 2, \ldots). \tag{2}$$

Convergence in mean. It is known that if the series (1) with respect to the system (2) converges in the metric of L^p, $1 < p \le 2$, or simply is the Fourier series of some function $f \in L^p$, then its coefficients have a power order of decrease:

$$c_n = o\left(\frac{1}{n^{1-\frac{1}{p}}}\right). \tag{3}$$

This follows from the classical Paley inequality concerning an arbitrary uniformly bounded ONS:

$$\sum_n |c_n(f)|^p n^{p-2} \le A_p \|f\|_p. \tag{4}$$

The exactness of the relation (3) is seen from a theorem of Hardy and Littlewood, according to which the series (1), $\phi_n = \cos nt$, with coefficients satisfying the condition $\sum |c_n|^p n^{p-2} < \infty$, converges in the L^p norm.

In the case $p = 1$ no inequalities of Paley's type exist. The question arises as to whether or not it is possible in this case to draw a conclusion about a definite rate of decrease of the coefficients of a mean-convergent series (or simply a Fourier series) with respect to a bounded system. For Fourier series (without requiring convergence) the answer is negative.

For example, the series $\sum c_n \cos nt$, where $\{c_n\}$ is a convex sequence, is always the Fourier series of some function $f \in L$ (see [10] Chap. I, § 30).

We shall show that for series that converge in mean the answer is positive: the coefficients must decrease faster than $\dfrac{1}{\ln n}$.

Theorem 1. *If the series* (1) *with respect to the ONS* $\{\phi_n\}$ *satisfies condition* (2) *and converges in the space* $L(X)$, *then*

$$c_n = o\left(\frac{1}{\ln n}\right). \tag{5}$$

It is convenient for the proof to consider an equivalent formulation: (i) *if the series* (1), *with* (2), *is bounded in* L, *then*

$$c_n = O\left(\frac{1}{\ln n}\right). \tag{6}$$

If the series (1) converges, then it is easy to find a sequence α_n growing sufficiently slowly to infinity so that the series $\sum \alpha_n c_n \phi_n$ will also converge. According to (i) we will have $\alpha_n c_n \ln n = O(1)$. That is, (5) is true. Thus from (i) follows the assertion of the theorem. The converse is easy to verify.

For the proof of assertion (i), we define

$$\lambda = \sup_{n,m} \left\| \sum_{k=n}^{m} c_k \phi_k \right\|_1 .$$

For each v set

$$n(v) = \min \{n; c_n \ln n > v\}$$

(if the set of numbers satisfying this inequality is nonempty).

Define also $n'(v) = \left[\dfrac{n(v)}{2}\right]$, $(v > 4c_1)$. Apply the fundamental inequality (§1) to the polynomial $\sum\limits_{n'+1}^{n} c_k \phi_k$. We will obtain

$$c_{n'(v)} \lambda \geq K(M) \frac{\sum\limits_{k=n'+1}^{n} c_k^2}{n-n'} \ln(n-n') \geq K(M) c_{n(v)}^2 \ln\left[n(v) - n'(v)\right].$$

From the definition of $n(v)$ it follows that

$$c_{n(v)} \ln n(v) > v \geq c_{n'(v)} \ln n'(v) \geq \frac{1}{\lambda} K(M) c_{n(v)}^2 \ln\left[n(v) - n'(v)\right] \ln n'(v);$$

that is,

$$c_{n(v)} \leq \frac{\lambda}{K} \frac{\ln n(v)}{\ln\left[n(v) - n'(v)\right] \ln n'(v)} \leq K_1(\lambda, M) \frac{1}{\ln n(v)} .$$

This, for sufficiently large v, contradicts the inequality $c_{n(v)} \ln n(v) > v$. Thus for some v we find that the number $n(v)$ is not defined. Therefore (6) is true, as was required.

In the case where the coefficients are convex, the exactness of Theorem 1 is well illustrated by the example of the trigonometric system: in this case, as is known (see [10] Chap. X, §2), condition (5) is not only necessary but is also sufficient for the convergence of the series $\sum c_n \cos nt$ in the space L. If the coefficients $\{c_n\}$ are only monotone then the sufficient conditions for the trigonometric system are somewhat stronger: $\sum \dfrac{c_n}{n} < \infty$. This is not by chance, as is seen from the following theorem.

Theorem 2. *For any bounded ONS it is possible to exhibit a series (1), divergent in the metric of L, whose coefficients satisfy condition (5).*

Nonetheless, Theorem 1 cannot be improved in class of all bounded systems:

Theorem 3. *For any sequence $\{c_n \downarrow\}$ satisfying condition (5), there can be found a uniformly bounded orthonormal system ϕ for which the series (1) converges in L.*

Thus condition (5), which is a necessary condition for the convergence in mean of the series (1) with respect to any bounded system, is at the same time sufficient—in the class of bounded systems, but not for any individual system in this class.

The proof of Theorem 3 depends on the following lemma.

Lemma. *Let there be given a natural number N and numbers*

$$1 \geq a_1 \geq a_2 \geq \cdots \geq a_n \geq \tfrac{1}{2}.$$

Then there exists a collection of trigonometric polynomials $\{t_i\}$ $(1 \leq i \leq n)$ *ON on* $[0,\pi]$ *(in which are present only harmonics* $> N$ *), possessing the following properties:*

(i) $\|t_i\|_\infty \leq 7$;

(ii) $\left\| \sum_1^n a_i t_i \right\|_1 < 1$;

(iii) $\left\| \sum_1^m a_i t_i \right\|_1 < C \ln(n+1)$ $(1 \leq m \leq n)$.

Fix a polynomial $\psi_0 = \sum\limits_{l=N+1}^{v} \gamma_l \cos l x$ satisfying the conditions $\|\psi_0\|_2 = 1$, $\|\psi_0\|_1 < \dfrac{1}{\sqrt{n}}$, $\|\psi_0\|_\infty < 2\sqrt{n}$. Suppose $\psi_j = \sqrt{\dfrac{2}{\pi}} \cos(v+j)x$ $(1 \leq j \leq n)$. Let $\alpha = \{\alpha_i\}$.

$$\alpha_i = \left(\sum_{j=1}^n a_j^2 \right)^{-1/2} a_i.$$

We shall consider the matrices $B^\alpha = \{b_{ij}\}$ $(0 \leq i,j \leq n)$:

$$b_{00} = 0, \quad b_{0j} = b_{j0} = \alpha_j, \quad b_{jj} = 1 - \alpha_j^2, \quad b_{ij} = -\alpha_i \alpha_j \quad (i \neq j, i,j > 0)$$

(in the particular case $\left\{ \alpha_i = \dfrac{1}{\sqrt{n}} \right\}$ they have already been used in the proof of

Theorem 6 § 2). It is easy to verify their orthogonality. Set $t_i = \sum\limits_{j=0}^{n} b_{ij} \psi_j$ $(1 \leq i \leq n)$. The system $\{t_i\}$ is orthonormal on $[0,\pi]$, and

$$\|t_i\|_\infty \leq \alpha_i \|\psi_0\|_\infty + \sum_{j=1}^n |b_{ij}| \leq \frac{2}{\sqrt{n}} \|\psi_0\|_\infty + 1 + \alpha_i \sum_{j=1}^n \alpha_j < 7;$$

$$\sum_{i=1}^n a_i t_i = \sqrt{\sum_{i=1}^n a_i^2} \cdot \sum_{i=1}^n \alpha_i t_i = \sqrt{\sum_1^n a_i^2} \cdot \psi_0,$$

whence follows (ii). Further,

$$\left\| \sum_{j=1}^m a_i t_i \right\|_1 = \left\| \sum_{i=1}^m a_i \left[\alpha_i \psi_0 + \psi_i - \alpha_i \sum_{j=1}^n \alpha_j \psi_j \right] \right\|_1,$$

$$\leq \sqrt{n} \|\psi_0\|_1 + \left\| \sum_{i=1}^m a_i \psi_i \right\|_1 + \sum_{i=1}^m \alpha_i^2 \left\| \sum_{j=1}^n a_j \psi_j \right\|_1.$$

Property (iii) follows from this and the inequality $\left\| \sum_1^k a_j \psi_j \right\| < C_1 \ln(n+1)$

$(k \leq n)$, which results from the Abel transformation and an estimate of the Dirichlet kernel.

Let there now be given a sequence $\{c_n \downarrow; \ c_n > 0\}$ satisfying condition (5). We fix the indices $\{v_k\}$ by the following conditions:

$$v_0 = 0; \qquad c_{v_k} \geq \tfrac{1}{2} c_{v_{k-1}+1}; \qquad c_{v_k+1} < \tfrac{1}{2} c_{v_{k-1}+1} \qquad (k \geq 1).$$

Successively, for each k, we apply the lemma, setting

$$a_i = \frac{c_{v_{k-1}+i}}{c_{v_{k-1}+1}} \qquad (1 \leq i \leq n_k \equiv v_k - v_{k-1})$$

and supposing N to be equal to the order of the highest harmonic occurring in the polynomials constructed on the previous $(k-1)$-th step. As a result we obtain a bounded orthonormal system $\{\phi_n\}$ $(1 \leq n < \infty)$ of trigonometric polynomials having the properties

$$\left\| \sum_{v_{k-1}+1}^{v_k} c_n \phi_n \right\|_1 < c_{v_{k-1}+1} \leq 2 c_{v_k};$$

$$\left\| \sum_{v_{k-1}+1}^{m} c_n \phi_n \right\|_1 \leq 2 C c_{v_k} \ln(n_k + 1) < C_3 c_{v_k} \ln(v_k + 1) \qquad (v_{k-1} < m \leq v_k).$$

Therefore for any $s < k$, $v_{k-1} < m \leq v_k$, we have

$$\left\| \sum_{n=v_s+1}^{m} c_n \phi_n \right\|_1 \leq \left\| \sum_{l=s+1}^{k-1} \sum_{n=v_{l-1}+1}^{v_l} c_n \phi_n \right\|_1 + \left\| \sum_{v_{k-1}+1}^{m} c_n \phi_n \right\|_1$$

$$\leq 2 \sum_{l=s+1}^{k-1} c_{v_l} + C_3 c_{v_k} \ln(v_k + 1).$$

Taking into account that the numbers c_{v_l} decrease in a geometric progression and using (5), we conclude that the series (1) converges in mean. By a simple addition to the construction it is possible to make the system in this theorem complete (closed in C).

We notice that Theorems 1 and 3 give necessary and sufficient conditions on a monotone sequence $\{c_n\}$ for series (1) to converge in mean for some bounded ONS. These conditions are (5). It would be interesting to investigate a similar question without assuming monotonicity. Questions of this type will be discussed at greater length in §3 of Chap. II.

Proof of Theorem 2. Using the inequality of §1, we can select a sequence of numbers fulfilling the conditions

$$\left\| \sum_{i=n_{k-1}+1}^{n_k} \phi_i \right\|_1 > \alpha \ln n_k \qquad (\alpha > 0); \tag{7}$$

$$\sum \frac{n_{k-1}}{\ln n_k} < \infty . \tag{8}$$

Letting $\Phi_k = \dfrac{1}{\sqrt{k \ln n_k}} \displaystyle\sum_{i=n_{k-1}+1}^{n_k} \phi_i$ we have, from (7), $\|\Phi_k\|_1 > \dfrac{\alpha}{\sqrt{k}}$. We shall make use of the following proposition.

Lemma (Orlicz [109]). *If*

$$\sup_{n:|\varepsilon_i|=1} \left\| \sum_{i=1}^{n} \varepsilon_i f_i \right\|_1 < \infty ,$$

then $\sum \|f_i\|_1^2 < \infty$.

Thus, there exists a sequence $k_s \uparrow$ such that

$$\limsup_{l \to \infty} \left\| \sum_{s=1}^{l} \Phi_{k_s} \right\|_1 = \infty . \qquad (9)$$

Set

$$c_n = \frac{1}{\sqrt{k_s \ln n_{k_s}}} \qquad (n_{k_s-1} < n \le n_{k_s}).$$

Clearly the numbers c_n are decreasing and satisfy condition (5). Further, we have

$$\sum_{n_{k_l}+1}^{n_{k_l}} c_n \phi_n = \sum_{s=2}^{l} \frac{1}{\sqrt{k_s \ln n_{k_s}}} \left[\sum_{j=k_s-1}^{k_s-1} \sum_{n=n_j+1}^{n_{j+1}} \phi_n \right]$$

$$= \sum_{s=2}^{l} \left[\frac{1}{\sqrt{k_s \ln n_{k_s}}} \sum_{j=k_s-1}^{k_s-2} \sum_{n=n_j+1}^{n_{j+1}} \phi_n + \Phi_{k_s} \right]; \qquad (10)$$

$$\left\| \sum_{s=2}^{l} \frac{1}{\sqrt{k_s \ln n_{k_s}}} \sum_{j=k_s-1}^{k_s-2} \sum_{n_j+1}^{n_{j+1}} \phi_n \right\|_1 \le \sum_{s=2}^{\infty} \frac{1}{\sqrt{k_s \ln n_{k_s}}} \sqrt{n_{k_s-1}} .$$

Therefore, keeping in mind (8) and (9), we conclude from (10) that the series (1) diverges in the metric of L.

Fourier coefficients. The sine series $\sum c_n \sin nx, c_n \searrow 0$ is a Fourier-Lebesgue series if and only if it converges in L. A necessary and sufficient condition for this is the condition $\sum \dfrac{c_n}{n} < \infty$. For cosine series the situation is different: the conditions on the Fourier coefficients are weaker than the conditions for convergence. For example, for convex sequences $\{c_n\}$, as has already been mentioned, the criterion for convergence is condition (5); at the same time, the series $\sum c_n \cos nx$, $c_n = o(1)$ with convex coefficients is always a Fourier series (Kolmogorov).

The condition on the second differences is essential here: as Sidon showed (see [10]), not every series $\sum c_n \cos nx, c_n \searrow 0$, is a Fourier series. This result was extended to the Walsh system by A. I. Rubinšteĭn, see [7].

We shall show that this result is true for any complete uniformly bounded orthonormal system [97]. Then it is possible to give an estimate of the rate of decrease of the numbers c_n that leaves little room for improvement [101].

Theorem 4. *For any uniformly bounded ONS $\{\phi_n\}$ that is complete in L there exists a series* (1), *with condition* (5), *such that this series is not a Fourier-Lebesgue series.*

It is not sufficient to assume here only one of the conditions of boundedness and completeness. The first assertion is confirmed by the example of the lacunary trigonometric system $\phi_n = \cos 2^n t$, with respect to which it is well known that every series $\sum c_n \phi_n$, $c_n = o(1)$ is a Fourier series. On the other hand the following result is true: there exists an ONS ϕ complete in L, possessing the property that every series $\sum c_n \phi_n$ with monotone bounded coefficients converges in L (see [92]).

For the proof of Theorem 4 we introduce the special Banach space V_ω of sequences. Let $\omega = \{0 < \omega(n) \to 0\}$. V_ω will denote the set of sequences $c = \{c_n\}$ satisfying the condition

$$|c_n| + \sum_{k=n}^{\infty} |c_{k+1} - c_k| = o(\omega(n)).$$

We introduce the norm

$$\|c\| = \sup_{n \geq 1} \frac{1}{\omega(n)} \left[|c_n| + \sum_{k=n}^{\infty} |c_{k+1} - c_k| \right].$$

It is not difficult to verify the following properties:

(i) V_ω is a Banach space.

(ii) $\forall c \in V_\omega$ $\|c^k - c\| = o(1)$, where $c^k = \{c_1, \ldots, c_k, 0, 0, \ldots\}$.

(iii) Each element $c \in V_\omega$ has the form $c = c' - c''$, where c' and c'' are monotone decreasing sequences, $c'_n = o(\omega(n))$, $c''_n = o(\omega(n))$.

We shall demonstrate only property (i).

Let $\lim_{p,q \to \infty} \|c^{(p)} - c^{(q)}\| = 0$, where $c^{(p)} = \{c_k^{(p)}\}$. Then clearly coordinatewise convergence takes place: $\lim_{p \to \infty} c_k^{(p)} = c_k$. Let $\|c^{(p)} - c^{(q)}\| < \varepsilon$ for $p, q > N(\varepsilon)$. Passing to the limit with respect to q in the inequality

$$|c_n^{(p)} - c_n^{(q)}| + \sum_{k=n}^{n'} |(c_{k+1}^{(p)} - c_{k+1}^{(q)}) - (c_k^{(p)} - c_k^{(q)})| < \varepsilon \omega(n) \qquad (n' > n),$$

we obtain

$$|c_n^{(p)} - c_n| + \sum_{k=n}^{n'} |(c_{k+1}^{(p)} - c_{k+1}) - (c_k^{(p)} - c_k)| \leq \varepsilon \omega(n).$$

Therefore in view of the arbitrariness of $n' > n$,

$$|c_n^{(p)} - c_n| + \sum_{k=n}^{\infty} |(c_{k+1}^{(p)} - c_{k+1}) - (c_k^{(p)} - c_k)| \leq \varepsilon \omega(n) \qquad (n = 1, 2, \ldots). \qquad (11)$$

Therefore

$$|c_n| + \sum_{k=n}^{\infty} |c_{k+1} - c_k| \leq \varepsilon \omega(n) + |c_n^{(p)}| + \sum_{k=n}^{\infty} |c_{k+1}^{(p)} - c_k^{(p)}| = \varepsilon \omega(n) + o(\omega(n)).$$

In view of the arbitrariness of $\varepsilon > 0$ we get $c = \{c_k\} \in V_\omega$. At the same time, inequality (11) gives $\|c^{(p)} - c\| \leq \varepsilon$ $(p > N)$.

For the proof of Theorem 4 we can now apply a device pointed out by Orlicz (see [55] Chap. VI, § 7), involving the Banach closed graph theorem. Assume that the system ϕ satisfies the conditions of the theorem, and that each series (1) satisfying (5) is a Fourier series. Then for each sequence $\{c_n' \downarrow\}$, $c_n' = o\left(\dfrac{1}{\ln n}\right)$, and so because of property (iii), each element $c = \{c_n\} \in V_\omega$, $\omega = \left\{\dfrac{1}{\ln n}\right\}$, is associated with some $f \in L$, $(f, \phi_n) = c_n$ ($\forall n$). Because of the completeness of the system, the vector f is uniquely defined. Therefore we have a linear operator $A: V_\omega \to L$ defined on the whole space V_ω. It is easy to see that this operator is closed, and consequently continuous. Taking into account (ii), we find that the series $\sum c_k \phi_k$ converges in the metric of L whenever $\{c_k\} \in V_\omega$. The latter contradicts Theorem 2.

We mention that for cosine series Theorem 4 was made more precise by L.A. Balashov: if the function $\lambda(n) \downarrow$ satisfies the condition $\sum \dfrac{\lambda(n)}{n} = \infty$, then there can be found a sequence $c_n \downarrow$, $c_n \leq \lambda(n)$ such that the series $\sum c_n \cos nx$ is not a Fourier series. At the same time the condition $\sum \dfrac{c_n}{n} < \infty$, $c_n \downarrow$, as is well known, guarantees even the convergence of the cosine series in L. We do not know whether the result of Balashov is true for any complete bounded system.

We mention that for the trigonometric system there are numerous investigations of the conditions on the Fourier coefficients when the requirement that they be convex or monotone sequences is somewhat weakened; see Boas [12], S.A. Telyakovskiĭ [148].

§ 4. Generalizations, Counterexamples, Problems

All the previous results involve uniformly bounded ONS. The question arises as to whether these results can be preserved for some wider class of systems. The question concerns weakening the condition of uniform boundedness, and concerns the consideration of nonorthogonal systems that are, in some certain sense, close to being orthogonal.

Several aspects of these problems are discussed briefly below.

Orthonormal systems with majorant. One of the most natural weakenings of uniform boundedness is the condition of boundedness at each point (or almost everywhere); that is,

$$\sup_n |\phi_n(x)| = \delta(x) < \infty \qquad (x \in X). \tag{1}$$

It is possible in addition to impose an integrability assumption on the majorant; for example, $\delta \in L^p$.

First of all, we notice that even in the case $\delta \in L^2$ (whence it follows that $\inf_n \|\phi_n\|_1 > 0$) it is not possible to give a nontrivial lower bound in the metric of L for the partial sums of series with respect to such systems (compare with § 1).

More precisely:

There exists an ONS $\{\phi_n\}$ that satisfies condition (1) with $\delta \in \bigcap_{p < \infty} L^p$ and that fulfills the inequality

$$\sup_n \left\| \sum_1^n \phi_k \right\|_1 < \infty. \tag{2}$$

Such an example can be obtained from the trigonometric system with the help of a unitary operator that arises from an appropriate change of variable. Let $t = t(x)$ be a strictly increasing absolutely continuous function of the interval $[0, \pi]$ onto itself whose inverse function is also absolutely continuous. Then, the operator

$$U : f(t) \mapsto f(t(x)) \sqrt{t'(x)} \tag{3}$$

is clearly unitary in $L^2[0, \pi]$. Let $\phi_k = U \psi_k$, where $\psi_k(t) = \sqrt{\dfrac{2}{\pi}} \sin kt$. Then

$$\delta(x) = O(\sqrt{t'(x)}), \quad \left\| \sum_1^n \phi_k \right\|_1 = \int_0^\pi \left| \sum_1^n \psi_k(t) \sqrt{x'(t)} \right| dt \leq C \int_0^\pi \frac{\sqrt{x'(t)} \, dt}{t}.$$

Considering $t(x) = \alpha \int_0^x \ln^4 \dfrac{\xi}{2\pi} \, d\xi$, where α is determined by the condition $t(\pi) = \pi$, we have $t'(x) = \alpha \ln^4 \dfrac{x}{2\pi}$.

Hence $\delta \in L^p$ ($\forall p < \infty$). Further, taking into account that $x(t) = O(t)$, we obtain $x'(t) = O\left(\ln^{-4} \dfrac{t}{2\pi} \right)$. Therefore (2) is true.

We mention that because of (2) each series $\sum c_k \phi_k, c_k \searrow 0$, converges in L. This shows that the results of §3 do not generalize to the class of systems we are examining.

The example given, however, does not mean that in this class it is possible to have bases in C, since in considering the Lebesgue functions we have a whole family of sets of coefficients $c_n = \phi_n(x)$, depending on the parameter x.

The following result shows that, without the assumption of integrability on the majorant, such bases do exist.

Theorem 1. *There exists an ONS $\{\phi_n\}$ satisfying the condition*

$$\phi_n(x) = o_x(1) \quad (\forall x), \tag{4}$$

with respect to which each continuous function expands as a uniformly convergent Fourier series.

We shall sketch the construction of this example; for details see [97]. We shall call a rearrangement $\{\psi_\nu\}$ of the Haar system normal if $\Delta[\psi_\nu] \supset \Delta[\psi_\mu]$ implies $\nu < \mu$ (where $\Delta[f]$ denotes the support of the function f). It is easy to prove the following.

Lemma. *Each normal rearrangement ψ of the Haar system satisfies the condition*

$$\sup_{n, x} L_n^\psi(x) < \infty. \tag{5}$$

The basic step in the construction is the following. Let $r, m, s, p, 1 \le m, s \le 2^r$. be given natural numbers. Let $T = T(r, m, s, p)$ denote the set $\{\psi_j\}$ $(0 \le j \le 2^p)$ of Haar functions determined by the conditions

$$\psi_0 = \chi_r^{(m)}; \quad \psi_j = \chi_{r+p}^{(j)}, \; j > 0: \quad \bigcup_{j=1}^{2^p} \varDelta[\psi_j] = \varDelta_r^{(s)} \equiv \left[\frac{s-1}{2^r}, \frac{s}{2^r}\right].$$

To this set we apply the transformation given by the matrices B_{2^p} (§ 2, relation (9)):

$$\phi_j(T, x) = \sum_{j=0}^{2^p} b_{ij}^{(2^p)} \psi_j. \tag{6}$$

Careful calculation leads to the inequality

$$\int_0^1 \left| \sum_{j=1}^l \phi_j(x) \phi_j(t) \right| dt < 8 \quad (x \in [0,1], l \le 2^p). \tag{7}$$

A simple (but cumbersome to state) induction then allows us to exhibit a normal rearrangement of the Haar system, consisting of the blocks $T_1', T_1, T_2', T_2, \dots$, where $T_k = T(r_k, m_k, s_k, l_k)$ has the form indicated and the blocks T_k' consist of functions with support in the interval $\varDelta_{r_k}^{(s_k)} = [\alpha_k, \beta_k]$. It is now possible to ensure that the following conditions are fulfilled:

(i) $p_k > r_k^2$;
(ii) $\alpha_k, \beta_k > 0$, $\lim \alpha_k = \lim \beta_k = 0$.
In each block T_k we pass to a new basis $\phi_k = \{\phi_i^{(k)}\}$ $(0 \le i \le 2^{p_k})$, according to (6). As a result we obtain an orthonormal system $\phi = \bigcup_k \{T_k', \phi_k\}$ that is equivalent to the Haar system. The lemma and inequality (7) ensure that its Lebesgue functions are uniformly bounded. It is easy to see that condition (i) gives the inequality $|\phi_i^{(k)}(x)| \le 2^{-\frac{r_k}{2}}$ $(x \notin \varDelta_{r_k}^{(s_k)})$, whence because of (ii) it follows that (4) holds.

The nature of the result changes if we require that $\delta \in L^2$.

Theorem 2. *If an ONS ϕ has a majorant $\delta \in L^2$, then*

$$\|L_n^\phi\|_x \ne O(1). \tag{8}$$

Indeed, the condition $\delta \in L^2$ implies that $\|\phi_n\|_1 > \alpha > 0$ $(\forall n)$. Therefore we obtain

$$\|L_n^\phi\|_x + \|L_{n-1}^\phi\|_x \ge \left\| \int_X |\phi_n(x) \phi_n(t)| dt \right\|_\infty > \alpha \|\phi_n\|_x.$$

The left and right sides of this inequality cannot be bounded simultaneously, by Theorem 1 § 1. Hence (8) follows.

The following questions remain open; positive answers seem more likely.
(i) Can (8) be preserved for complete systems such that $\delta \in L^p, p < 2$?
(ii) Is it possible in Theorem 2 to assert that $L_n(x) \ne O(1)$ for some x (that is, there is not only not uniform convergence, but also divergence of the Fourier series of a continuous function)?

However, a simultaneous strengthening of Theorem 2 in both the directions indicated is not possible.

There exists an ONS $\{\psi_n(x)\}$ with $\delta \in \bigcap_{p<2} L^p$, such that any continuous func-
tion expands as an everywhere convergent Fourier series.

It is possible, with the help of the operator U defined by equality (3), to obtain such a system from the system $\{\phi_n(t)\}$ constructed in Theorem 1. In this case the function $x = x(t)$ that implements the change of variable must go sufficiently rapidly to zero.

We mention also the following proposition, formally somewhat strengthening Theorems 2 and 9 of § 2.

If an ONS $\{\phi_n\}$ forms a basis in C (or in L), then it satisfies the relations

$$\lim\inf\|\phi_n\|_p = 0 \quad (p < 2); \quad \lim\sup\|\phi_n\|_p = \infty \quad (p > 2). \tag{9}$$

For $p = 1$, this is contained in the proof of Theorem 2. The general case follows from the case $p = 1$ and the equality $\|\phi_n\|_2 = 1$, because of the convexity of the function $\alpha_f(p) = \ln \int_X |f|^p dt$.

The terms lim inf and lim sup in (9) cannot be interchanged: there exist examples that are orthonormal bases in C and contain infinite uniformly bounded subsystems (see the proof of Theorem 8 § 2).

Slowly growing systems. Let $M_n \nearrow \infty$ be a given sequence. We shall consider the class of orthonormal systems satisfying the condition

$$\|\phi_n\|_\gamma = O(M_n). \tag{10}$$

Are the basic results of § 2 preserved for sufficiently slowly growing numbers M_n? It turns out that the answer essentially depends on whether the question concerns uniform convergence of Fourier series or convergence everywhere.

The following proposition holds.

For any sequence $M_n \nearrow \infty$, there exists an ONS ϕ satisfying condition (10), with respect to which each continuous function expands as a Fourier series converging to it everywhere.

Such an example can be constructed without difficulty. It is sufficient to choose for each k a block of Haar functions $\{\chi_m^s\}$, $k \leq m < k + p_k$, where for each m the index s takes all values for which $\Delta[\chi_m^{(s)}] \subset \left[\dfrac{1}{2^k}, \dfrac{1}{2^{k-1}}\right]$. In each block it is possible to pass to a new orthonormal basis $\{\phi_i^{(k)}\}$ of the Walsh type so that the condition $\|\phi_i^{(k)}\|_\infty = \sqrt{2^k}$ is satisfied. It is easy to find numbers p_k such that the system $\phi = \bigcup\{\phi_i^{(k)}\}$ satisfies condition (10). It remains to place into sufficiently rare places in the correct order all the Haar functions that do not take part in this construction.

At the same time, in reference to uniform convergence, the result is preserved to a certain extent when passing from the class of bounded systems to the class of slowly growing systems.

If an ONS satisfies condition (10), where the numbers M_n grow sufficiently slowly, for example $M_n = o(\sqrt{\ln n})$, then condition (8) is satisfied.

Indeed, the inequality of § 1 gives

$$\max_{1 \le m \le n} \int_X \left| \sum_{k=1}^m \phi_k(x)\phi_k(t) \right| dt \ge \frac{K_0}{M_n^2} \frac{\sum_{1}^{n} \phi_k^2(x)}{n} \ln n \quad (\forall x).$$

Hence, after integrating, we obtain

$$\max_{1 \le m \le n} \|L_n\|_\gamma \ge \frac{K_0}{\mu X} \frac{\ln n}{M_n^2} \to \infty \quad (n \to \infty).$$

An interesting question is

(iii) to estimate the maximum possible order of growth of M_n in this statement. It is possible that the result is preserved for $M_n = O(n^{\frac{1}{2}-\varepsilon})$. At the same time for the Haar system we have $\|\chi_n\|_\infty = O(\sqrt{n})$. There exists an intrinsic connection between this question and problem (i) of the preceding subsection.

Nonorthogonal bases. In conclusion we offer a few words on the extension of these results to nonorthogonal systems.

Upon examination of the proof of the fundamental inequality, it is readily noticed that the orthogonality of the system $\{\phi_i\}$ is used only in the form of the following inequality, which is true for any n and $\{c_i\}$:

$$\left\| \sum_{i=1}^n c_i \phi_i \right\|_2 \ge K_1 \left(\sum_{i=1}^n c_i^2 \right)^{1/2}, \quad K_1 > 0.$$

Systems having this property are called Bessel systems.

Let the systems $\{\psi_i\}$ and $\{\phi_i\}$ form a biorthogonal system in L^2; that is, $(\phi_i, \psi_j) = \delta_{ij}$. Then each function $f \in L^2$ is associated with a Fourier series $f \sim \sum c_i(f)\psi_i$, $c_i = (f, \phi_i)$. The Lebesgue functions in this case have the form $L_n(x) = \int_X \left| \sum_{i=1}^n \phi_i(t)\psi_i(x) \right| dt$. It is known that if $\{\phi_i\}$ is Bessel, then $\{\psi_i\}$ is a Hilbert system and vice versa. A system $\{\psi_i\}$ is called Hilbert if the inequality $\|\sum c_i \psi_i\|_2 \le K_2 (\sum c_i^2)^{1/2}$ holds. The above-mentioned extension of the fundamental inequality leads to the following proposition, which generalizes Theorem 2 § 2.

If $\{\psi_i\}$ is a normalized Hilbert system in L^2 that forms a basis in the space C then $\|\phi_i\|_\infty \ne O(1)$.

The following hypothesis also seems probable.

(iv) If the system $\{\psi_i\}$ forms a basis simultaneously in C and in L^2, then $\|\psi_i\|_\gamma \ne O(\|\psi_i\|_2)$.

At the same time, it is not possible to quite get rid of some type of orthogonality condition: there exists a basis $\{\psi_i\}$ in the space C that satisfies the condition $\inf \|\psi_i\|_2 > 0$; $\sup \|\psi_i\|_\gamma < \infty$ (Meletidi [73]).

Such an example is constructed from the classical Schauder basis $\{f_k^0\}$ (see below § 5) in the following manner: it is not difficult to find numbers $\varepsilon_k = \pm 1$ so that the condition $1 \le \sum_{k=1}^n \varepsilon_k f_k^0(x) \le 3$, $x \in [0,1]$ is satisfied. Then the system $\psi_n = \sum_{k=1}^n g_k$, $g_k = \varepsilon_k f_k$, forms the desired basis in $C[0,1]$. Indeed, if $c_k(f)$ are the

coefficients of the vector f with respect to the basis $\{g_k\}$, then after the Abel transformation $\sum_1^n c_k g_k = \sum_1^n (c_k - c_{k+1}) \psi_k + c_{n+1} \psi_n$ we see that f expands as the series
$\sum_1^\infty d_k \psi_k$, $d_k = c_k - c_{k+1}$. The uniqueness of the expansion follows from the existence of a system of functionals dual to $\{\psi_k\}: \psi_k^* = g_k^* - g_{k+1}^*$ where the system
$\{g_k^*\}$ is dual to $\{g_k\}$.

The following two hypotheses were reported to the author by V. I. Gurariĭ.

(v) If $\{\psi_n\}$ is a normalized basis in C then for any sequence $\alpha_k \to 0$ there
exists a vector f and a sequence of numbers $\{n_k\}$ such that $|c_{n_k}(f)| > \alpha_k$.

(vi) If $\{\psi_n\}$ is a normalized basis in L then $\|\psi_n\|_\infty \neq O(1)$. The last problem
arose in connection with Theorem 9 § 2.

§ 5. The Stability of the Orthogonalization Operator

One of the general methods of constructing orthogonal systems is the classical
Schmidt orthogonalization process. If $f = \{f_k\}$ is a linearly independent system
of vectors in a Hilbert space H, then by $\psi = Sf$ is meant the ONS obtained by
orthogonalizing this system.

The application of the operator S to the system of functions $1, t, t^2, \ldots$ in the
space L_μ^2 leads to a classical system of orthogonal polynomials. With the help
of the operator S, Franklin was first able to construct an ONS forming a basis
in the space $C[0,1]$. Here the orthogonalization process was applied to the
Schauder system f^0.

As is well known, Schauder indicated a certain class of bases f^ω in the space C.
Namely, if $\omega = \{\omega_k\}$ is a sequence of distinct numbers $(\omega_1 = 0, \omega_2 = 1)$ that is
dense in $[0,1]$, then we define $f_1^\omega(t) \equiv 1$, $f_2^\omega(t) = t$, and further, for each $k > 2$,
f_k^ω is defined to be the piecewise linear function with corners $\{\omega_i\}_1^k$, and
$f_k^\omega(\omega_k) = 1$, $f_k^\omega(\omega_i) = 0$ $(1 \leq i < k)$. The resulting system f^ω forms a basis in $C[0,1]$.
In particular, the standard Schauder basis f^0 arises from the sequence
$\omega = \{0, 1, 1/2, 1/4, 3/4, 1/8, 3/8, \ldots\}$. (As is observed in [43], this basis was first
pointed out by Faber).

The Franklin system $\psi^0 = Sf^0$ is a basis in $C[0,1]$. As Ciesielski has shown,
this result extends to the systems $\psi^\omega = Sf^\omega$. This author also investigated in
detail [20] the properties of the Franklin system, which in many respects are
analogous to properties of the Haar system.

The present section consists of the formulation of results on the stability of
the operator S, and several applications to the construction of ON bases in
function spaces. The first statement of the problem can be illustrated by the
following example. Let $\mathscr{D}$ be an everywhere dense linear subspace of the space C.
The problem consists of constructing an ONS $\psi = \{\psi_k\}$, $\psi_k \in \mathscr{D}$, that forms a
basis in this space.

In the case where the orthogonality of the system is not required, the problem
is easily solved as a result of the well-known consideration of the stability of bases.
It is sufficient to use the following proposition.

If $f = \{f_k\}$ is a normalized basis in the Banach space B and satisfies the inequality $\sum \|\tilde{f}_k - f_k\| < \dfrac{1}{2K(f)}$ then the perturbed system $\tilde{f}$ is also a basis.

This theorem, due to M. Kreĭn, Milman and Rutman (see [47]), is one of the results describing the property of stability of bases relative to small perturbations. There are other well-known propositions of this type (see [41, 77]). The general method of obtaining them, apparently going back to Paley and Wiener [165], consists of proving that the transformation $f_k \to \tilde{f}_k = f_k + \Delta_k$ for small perturbations Δ_k yields a linear homeomorphism of the space onto itself. For this in turn it is sufficient to show that the operator $A: f_k \mapsto \Delta_k$ has norm < 1, after which we use the expansion of the operator $I - A$ as a Neumann series. Another variant of this method is to show that the operator A is compact and then use the Fredholm alternative. The theorem of N. K. Bary on the quadratic stability of bases in Hilbert space is proven in this way (see [47]).

For the construction of orthogonal bases $\psi, \psi_k \in \mathscr{D}$, the natural approach consists of approximating the Schauder basis f^0 by elements of the set $\mathscr{D}$ and then orthogonalizing. Taking into account that the system Sf^0 is a basis, it can be expected that this property is maintained even after small perturbations. Such an approach was used by K. M. Shaĭdukov [119] who first constructed an ON basis in C consisting of algebraic polynomials. However in the general case one encounters some surprises when carrying out this approach.

Namely, Szlenk [131] showed that for any sequence $\varepsilon = \{\varepsilon_k > 0\}$ it is possible to exhibit a basis f^ω and a perturbed basis $\tilde{f}$, $\|f_k^\omega - \tilde{f}_k\| < \varepsilon_k$ ($k = 1, 2, \ldots$), for which the ONS $\tilde{\psi} = S\tilde{f}$ no longer forms a basis in C.

It is true that the basis f^ω that is perturbed in the Szlenk construction depends on ε, so that this result shows only that the operator S is not uniformly stable on the entire class of bases f^ω. However it was soon discovered [104] that there is in fact a stronger local instability. More precisely, fixing an arbitrary basis f^ω, we can by arbitrarily small pertubations of the elements change it into a system $\tilde{f}$ that is not a basis after orthogonalization. In fact it is sufficient to apply the perturbation to only one (any one) vector of the system f^ω, leaving all the others unchanged.

The geometric nature of such instability is made clear in the formulation below of Theorem 1, which gives criteria for the stability of the operator S.

We shall examine the following more general situation. Let B be a separable Banach space in which, besides the initial norm $\|x\|$, there is a continuous inner product (x, y); that is, B is dense in a Hilbert space H.

Henceforth the orthonormality of a system will be understood in the Hilbert space sense, and a system will be called a basis if it is a basis in the Banach space B. The norm in H will be denoted by $|x|$.

We shall say that the pair $B \subset H$ satisfies condition γ if

(a) the norms $|x|$ and $\|x\|$ are not equivalent; that is, $B \neq H$;

(b) there exists an ONS $\{g_n\}$ bounded in B.

This condition is fulfilled, for example, in the case $H = L^2$, $B = C$ or L^p ($2 < p < \infty$).

An ε-neighborhood $U(\varepsilon, f)$ of the system f will be the set of all systems $\tilde{f}$ for which $\|\tilde{f}_k - f_k\| < \varepsilon_k \ \forall k \ (0 < \varepsilon_k$ are given).

Let f be a system such that Sf is a basis in B. We shall call f a *point of stability* of the operator S if there exists a neighborhood $U(\varepsilon, f)$, such that $\tilde{f} \in U \Rightarrow S\tilde{f}$ is a basis.

The following theorem is true.

Theorem 1. *In order that the system f be a point of stability of the operator S it is sufficient, and with condition γ also necessary, that the following conditions be satisfied:*

(i) *f is minimal in H;*

(ii) $\sup_n \|P_n^{(k)} f_k\| < \infty \ (\forall k)$, *where $P_n^{(k)}$ is the orthogonal projection of the space B onto the subspace F_n^k generated by the vectors $\{f_i\} \ (1 \leq i \leq n, i \neq k)$.*

Thus, if condition (i) is violated, that is, some vector f_k belongs to the closure (in H) of the linear span of the other vectors, then when condition γ is fulfilled, the operator S is unstable relative to small perturbations of only the vector f_k. This remark applies also to condition (ii).

Theorem 1 gives a classification of the points of instability: they can be of two types, according to which of the conditions of the theorem is violated. We shall illustrate this in the case $H = L^2$, $B = C$.

Examples of points of instability of the first type are the Schauder systems f^ω. It is readily noticed that each vector of the system f^ω is approximated in L^2 arbitrarily closely by linear combinations of the other vectors; therefore a small perturbation of any vector of the Schauder basis leads to instability. It should not be thought, however, that this is a result of the loss of linear independence: the perturbations are small in the metric of C, in which the Schauder system forms a basis and in particular is minimal.

The violation of condition (i) means that the system f is essentially non-orthogonal. At the same time points of instability of the second type can be located close to ON bases. For example, the following proposition is true: *for any ON basis ϕ in the space C and any $\delta > 0$, there can be exhibited a basis f, $\sum \|f_k - \phi_k\| < \delta$, which is a point of instability (of the second type) of the operator S.*

The proof depends on Theorem 1 and Theorem 2 § 2. At the same time it follows from Theorem 1 that

Theorem 2. *Every ON basis of the space B is a point of stability of the operator S.*

That conditions (i) and (ii) are fulfilled is in this case obvious. The answer to the question given above follows from Theorem 2.

If in the space B there exists a basis ϕ, ON relative to H, then such a basis can be constructed from the elements of a given dense linear subspace $\mathscr{D}$.

For example, in case of the space C the method of constructing such a basis consists of approximating the Franklin system by elements of the set $\mathscr{D}$ and then orthogonalizing. We mention that this approach, which was pointed out in the work [104], was used by Z. A. Chanturiya [19] to get an upper bound for the possible rate of increase of the degree of the polynomials in an orthogonal basis in C.

A more general example is the space C_n^k of functions of smoothness k on the n-dimensional cube. Not long ago Z. Ciesielski [21] obtained a positive solution for the problem of the existence of a basis in this space; the basis he constructed is orthogonal in L^2. Hence the above-mentioned result permits us to conclude the existence in this space of an ON basis consisting of algebraic polynomials.

The importance of the condition of continuity of the inner product can be illustrated by the following example. We fix a function $f_0 \in L^2$, $f_0 \notin L^{\frac{p}{p-1}}$ $(p<2)$. Let $\mathcal{D}$ be the set of all smooth functions orthogonal to f_0. From the Hahn-Banach theorem it is easy to see that $\mathcal{D}$ is dense in L^p. However it is not possible to exhibit an ONS $\{\phi_n \in \mathcal{D}\}$ that forms a basis in L^p. Indeed, if it were possible we would have $f_0 = \sum d_n \phi_n$ (in L^p), whence because of the condition $\phi_n \in \mathcal{D} \subset L^{\frac{p}{p-1}}$, we would obtain $d_n = (f_0, \phi_n) = 0$, which is a contradiction.

The proof of the assertions formulated above is contained in [104] (notice that condition γ, not introduced so obviously there, is actually included in Lemma 3 of the work cited).

We also mention the following fact established in passing in the proof of Theorem 1, and independently obtained in [46].

If a system of vectors $\{f_k\}$ is closed in the Banach space B then there exists a neighborhood $U(\varepsilon; f)$ such that every system $\tilde{f} \in U$ is closed.

This means that the property of a system being closed is stable. We mention that the stability of various properties of orthogonal systems was investigated in detail by N. K. Bary (see [47]). A noteworthy point of the lemma above consists of the possibility of a nonminimal system f such that the approach involving the examination of the operator $f_k \to \tilde{f}_k$ (see above) is inapplicable.

In a neighborhood of a fixed point of the operator S there is a stronger quadratic stability: for small perturbations of elements of the ON basis ϕ, the system $\tilde{\psi} = S\tilde{\phi}$ obtained by orthogonalization turns out to be close to the initial system in the sense of the distance

$$\rho(\phi, \psi) = (\sum \|\psi_k - \phi_k\|^2)^{1/2}.$$

More precisely, the following is true:

Theorem 3 (see [104]). *Let ϕ be an ON basis in the space B. Then for any number $\delta > 0$ there exists a neighborhood $U(\varepsilon; \phi)$ such that if $\tilde{\phi} \in U$ then the inequality $\rho(\tilde{\psi}, \phi) < \delta$ holds.*

It is easy to show that for sufficiently small δ this inequality ensures that the system $\tilde{\psi}$ is a basis.

This result gives a way of approximating an ON basis by means of similar bases consisting of elements of a given dense linear subspace. The method of approximation indicated is in a certain sense extremal. It is possible to show, for example (see [104]), that *there exists an ONS, complete in $L^2[0,1]$, of piecewise continuous functions $\{\phi_n\}$ such that any ONS of continuous functions $\{\psi_n\}$ satisfies the condition $\sum \|\phi_n - \psi_n\|_2^p = \infty$ $(\forall p < 2)$.*

We mention also that the condition that ϕ be a basis is essential for the proximity of the systems $\tilde{\psi} = S\,\tilde{\phi}$ and ϕ. The following assertion is true.

If ϕ is a closed ONS that is not a basis in B then after some arbitrarily small perturbation Δ_1 the relation

$$\|\tilde{\psi}_k - \phi_k\| \neq O(1)$$

is satisfied, where $\tilde{\phi} = \{\phi_1 + \Delta_1, \phi_2, \phi_3, \ldots\}$.

Chapter II. Convergence Almost Everywhere; Conditions on the Coefficients

With the emergence of Lebesgue measure theory, it was hoped that the points of divergence of Fourier series could be characterized, in terms of this theory, as being exceptional. The following two fundamental results were therefore a great surprise.

1. (D. E. Menshov, 1923). There exists an ONS $\{\phi_n\}$ such that some series

$$\sum c_n \phi_n \tag{1}$$

whose coefficients satisfy the condition

$$\sum c_n^2 < \infty \tag{2}$$

diverges almost everywhere.

2. (A. N. Kolmogorov, 1927). There exists a function $f \in L$ whose Fourier series with respect to the trigonometric system diverges almost everywhere (and even everywhere).

These facts were used as a point of departure for many subsequent investigations. Some of them will be presented in this chapter. Other results relating to these questions will be found in Chap. III (§§ 2 and 5) and in Chap. IV.

In this chapter we shall study conditions on the coefficients for convergence almost everywhere. The general problem is to describe, for a given ONS ϕ, the class $\mathfrak{S}(\phi)$ of sequences of coefficients $\{c_n\}$ for which the series (1) converges almost everywhere.

Only in rare instances is there an effective solution to this problem. One of these exceptions, for example, is the Rademacher system $\mathfrak{r}$, for which $\mathfrak{S}(\mathfrak{r}) = l_2$ (Rademacher, Kolmogorov). Such a result holds for lacunary trigonometric systems (Kolmogorov, Zygmund) and for other systems of the same type (see the survey [39]).

As a rule, the description of the class $\mathfrak{S}(\phi)$ for an individual system is a very difficult task. It should be sufficient to note that Lusin's hypothesis, the containment $\mathfrak{S}(\tau) \supset l_2$ (τ being the trigonometric system), remained unproved for half a century, until the work of Carleson.

It is therefore of interest to investigate the intersection (or the union) of the classes $\mathfrak{S}(\phi)$ for various sets of ON systems. Special attention has been given to the class

$$\mathfrak{S}_\Omega = \bigcap_{\phi \in \Omega} \mathfrak{S}(\phi)$$

where Ω is the set of all ONS on X. The results obtained for this class are presented in § 1.

§ 1. The Class $\mathfrak{S}_\Omega$

Weyl multipliers; the classical results. The theorem of Menshov stated above (for a simple proof, see Chap. III; § 2, Lemma 1) means that all ONS fall into two classes: *systems of convergence,* for which $\mathfrak{S}(\phi) \supset l_2$, and (all others) *systems of divergence.* The existence of the latter makes us interested in conditions like Weyl's for the convergence of Fourier series.

A sequence $\omega(n)\uparrow\infty$ is called a Weyl multiplier for the system ϕ if the condition

$$\sum c_n^2 \omega(n) < \infty \tag{3}$$

implies the convergence of series (1) almost everywhere; that is, if the weighted space l_2^ω is contained in $\mathfrak{S}(\phi)$.

Condition (3) for convergence almost everywhere has been investigated by many authors, beginning with H. Weyl [164]. The conclusive result was proved by D. E. Menshov and H. Rademacher:

the sequence $\omega(n) = \ln^2 n$ is a Weyl multiplier for any ONS.

The importance of this theorem is due to the fact that no further advance is possible in the general case. D. E. Menshov *constructed an ONS ϕ for which the sequence $\{\ln^2 n\}$ is an exact Weyl multiplier* (that is, no sequence $\omega(n) = o(\ln^2 n)$ is a Weyl multiplier for the system ϕ).

These theorems are based on the following two fundamental lemmas (for proof, see [55]).

Lemma 1 (Menshov-Rademacher). *The following inequality holds for any ONS ϕ and for any collection of numbers $\{c_k\}_1^n$:*

$$\int_X \left[\max_{1 \le l \le n} \left| \sum_{i=1}^l c_i \phi_i(x) \right|^2 \right] d\mu \le K \sum_{i=1}^n c_i^2 \ln^2 n \quad (n > 1), \tag{4}$$

where K is an absolute constant.

Lemma 2 (Menshov). *For any n there exists a collection of functions $\{\phi_i\}$ $(1 \le i \le n)$, orthonormal on $[0,1]$, satisfying the inequality*

$$\max_{1 \le l \le n} \left| \sum_{i=1}^l \phi_i(x) \right|^2 > K_1 n \ln^2 n; \quad (x \in E, \mu E > \tfrac{1}{5}), \quad K_1 > 0.$$

Menshov's method was developed further by K. Tandori to obtain more general results.

Tandori's results. For any given collection of numbers $\{c_1, \ldots, c_n\}$, define

$$I(c_1, \ldots, c_n) = \sup_{\phi \in \Omega} \int_X \max_{1 \le l \le n} \left| \sum_{i=1}^l c_i \phi_i(x) \right|^2 dx; \quad X = [0,1]. \tag{5}$$

The following properties are readily verified (see [142]):

(i) $I(c_1, \ldots, c_n) \ge \sum_1^n c_i^2$;

(ii) $I(c_1, \ldots, c_n) \le I(c_1, \ldots, c_n, c_{n+1})$;

(iii) $I^{1/2}(c_1+d_1,\ldots,c_n+d_n) \le I^{1/2}(c_1,\ldots,c_n) + I^{1/2}(d_1,\ldots,d_n)$;

(iv) $I(c_1,\ldots,c_n,d_1,\ldots,d_m) \ge I(c_1,\ldots,c_n) + I(d_1,\ldots,d_m)$.

For proof of the last assertion it suffices to pick ONS ϕ and ϕ' for which the integral in (5) comes within ε of the supremum for the collections c and d respectively, and to consider the system ψ:

$$\psi_i(x) = \begin{cases} \sqrt{2}\,\phi_i(2x) & (0 \le x \le \tfrac{1}{2}) \\ 0 & (\tfrac{1}{2} < x \le 1) \end{cases}, \quad 1 \le i \le n;$$

$$\psi_{n+j}(x) = \begin{cases} 0, & x \le \tfrac{1}{2} \\ \sqrt{2}\,\phi_j'(2x-1), & x > \tfrac{1}{2} \end{cases}, \quad 1 \le j \le m.$$

(v) $I(c_1,\ldots,c_n) \le I(d_1,\ldots,d_n)$ if $|c_i| \le |d_i|$ $(1 \le i \le n)$.

The proof is analogous to the preceding one; assume $d_i \ne 0$ and consider the ONS ψ:

$$\psi_i(x) = \begin{cases} \sqrt{2}\dfrac{c_i}{d_i}\,\phi_i(2x), & 0 \le x \le \tfrac{1}{2} \\[2mm] \sqrt{2}\sqrt{1 - \dfrac{c_i^2}{d_i^2}}\,\phi_i(2x-1), & \tfrac{1}{2} < x \le 1 \end{cases}.$$

The next proposition is essential for what follows.

Lemma 3. *For any collection* $c_1,\ldots,c_n$, $\sum |c_i| > 0$, *it is possible to exhibit an ON collection of step functions* $\{\psi_i\}$ $(1 \le i \le n)$ *satisfying the relation*

$$\mu\left\{x;\ \max_{1 \le l \le n}\left|\sum_{i=1}^{l} c_i\psi_i(x)\right| > \tfrac{1}{2}\right\} > \tfrac{1}{2}\min\{I(c_1,\ldots,c_n),1\}. \tag{6}$$

Based on (5), fix an ONS $\{\phi_i\}$ $(1 \le i \le n)$ for which

$$\alpha = \|\delta\|_2^2 > \tfrac{1}{2}I(c_1,\ldots,c_n), \quad \delta(t) = \max_{1 \le l \le n}\left|\sum_{i=1}^{l} c_i\phi_i(t)\right|.$$

Without loss of generality we can assume that

$$\delta(t) > 0 \ (\forall t); \quad \alpha < 1.$$

Then the function $x(t) = \int_0^t \delta^2(u)\,du$ defines a change of variable $[0,1] \to [0,\alpha]$ which is absolutely continuous in both directions. Let U be the unitary operator corresponding to the inverse function $t = t(x)$ by formula (3) of § 4, Chap. I. Let $\tilde{\phi}_i(x) = (U\phi_i)(x) \equiv \phi_i(t(x))\sqrt{t'(x)}$. For $x \in (\alpha, 1]$ set $\tilde{\phi}_i(x) = 0$. We have

$$\tilde{\delta}(x) = \max_{1 \le l \le n}\left|\sum_{i=1}^{l} c_i\tilde{\phi}_i(x)\right| = \delta(t)\sqrt{t'(x)} = 1$$

almost everywhere on $[0,\alpha]$. Approximating $\tilde{\phi}_i$ in L^2 by ON step functions, we obtain a system ψ with the required properties.

Now assign to each sequence $\{c_i\}$ the quantity

$$\|\{c_i\}\| = \lim_{n \to \infty} I^{1/2}(c_1,\ldots,c_n). \tag{7}$$

(The existence of the limit, finite or infinite, is implied by (ii).) This function has all the properties of a norm. (The triangle inequality follows from (iii).)

Theorem 1 [142]. *The class $\mathfrak{S}_\Omega$ coincides with the set of all sequences $c = \{c_i\}$ for which the norm (7) is finite.*

Indeed, suppose $\|c\| < \infty$. Then property (iv) implies for any sequence of indices $n_k \uparrow$ the relation

$$\sum_k I(c_{n_k+1}, \ldots, c_{n_{k+1}}) < \infty. \tag{8}$$

Let ϕ be a fixed ONS. On the basis of (i), choose indices $\{n_k\}$ such that the partial sums $S_{n_k}(x)$ of the series (1) converge almost everywhere. With the notation

$$\delta_k = \max_{n_k < l \leq n_{k+1}} \left| \sum_{i=n_k+1}^{l} c_i \phi_i(x) \right|, \text{ we have, by (8), } \sum \|\delta_k\|_2^2 < \infty, \text{ so that } \delta_k(x) = o_x(1)$$

almost everywhere, whence follows the convergence of series (1). This means that $c \in \mathfrak{S}_\Omega$.

Now suppose $\|c\| = \infty$. Then by property (iii) it is possible to exhibit a sequence of indices $\{n_k \uparrow\}$ that satisfies the inequality

$$I(c_{n_k+1}, \ldots, c_{n_{k+1}}) > 1 \quad (\forall k).$$

Using Lemma 3 and the usual technique associated with stochastically independent sets, we can construct an ONS ϕ for which the series (1) diverges almost everywhere. Specifically, for each k, Lemma 3 gives an ONS of step functions $\{\psi_i\}$ $(n_k < i \leq n_{k+1})$ such that

$$\mu\left\{x; \max_{n_k < l \leq n_{k+1}} \left| \sum_{i=n_k+1}^{l} c_i \psi_i(x) \right| > \frac{1}{2}\right\} > \frac{1}{2}. \tag{9}$$

We proceed to construct inductively the ONS of step functions $\{\phi_i\}$ $(1 \leq i < \infty)$. Assuming the functions ϕ_i for $i \leq n_k$ have already been constructed, denote by ρ_m the intervals on which they are all constant; $\bigcup \overline{\rho_m} = X$. Let $\rho_m^{(j)}$ $(j=0,1)$ be respectively the left and right halves of the interval ρ_m. Set

$$\phi_i(t) = (-1)^j \psi_i(x_m^{(j)}(t)), \; t \in \rho_m^{(j)}, \quad \text{where} \quad x_m^{(j)}: \rho_m^{(j)} \to (0,1)$$

is a linear change of variable. This provides the ONS ϕ, and by virtue of (9) we have

$$\mu E_k > \frac{1}{2}, \quad E_k = \left\{x; \max_{n_k < l \leq n_{k+1}} \left| \sum_{i=n_k+1}^{l} c_i \phi_i(x) \right| > \frac{1}{2}\right\},$$

and the sets E_k are seen to be independent. Series (1) clearly diverges everywhere on the set $E = \limsup E_k$, which has full measure. Theorem 1 is now proved. Also true is the following result (see [145]):

the class $\mathfrak{S}_\Omega$ is a separable reflexive Banach space with respect to the norm (7).

(The linear operations are understood in the usual sense.) From the general properties of the norm (7) the following proposition is deduced [142]: *if $\{c_n\} \in \mathfrak{S}_\Omega$ $(\{c_n\} \notin \mathfrak{S}_\Omega)$, then there exists a sequence $\lambda_n \uparrow \infty$ (respectively $\lambda_n \downarrow 0$) such that $\{\lambda_n c_n\} \in \mathfrak{S}_\Omega$ (respectively $\{\lambda_n c_n\} \notin \mathfrak{S}_\Omega$).*

In order to make the description of the class $\mathfrak{S}_\Omega$ given by Theorem 1 effective, one must know how to compute the norm (7) explicitly. This problem has not been solved in the general case. Inequality (4) implies the following upper bound:

$$\|\{c_n\}\| \le K_2 \left(c_1^2 + \sum_{n=2}^{\gamma} c_n^2 \ln^2 n \right)^{1/2}. \tag{10}$$

Indeed, for any ν, $1 \le l \le 2^\nu$, we have the following inequality for the partial sums S_l of series (1):

$$|S_l(x)| \le |S_{2^\nu}(x)| + \left\{ \sum_{n=1}^{\nu} |S_{2^n}(x) - S_{2^{n-1}}(x)|^2 \right\}^{1/2}$$

$$+ \left\{ \sum_{n=2}^{\nu} \max_{2^{n-1} < l \le 2^n} \left| \sum_{i=2^{n-1}+1}^{l} c_i \phi_i(x) \right|^2 \right\}^{1/2}.$$

Therefore

$$\int_X \max_{1 \le l \le 2^\nu} |S_l(x)|^2 \, dx \le K_3 \left(\sum_{i=1}^{2^\nu} c_i^2 + \sum_{n=1}^{\nu} \sum_{i=2^{n-1}+1}^{2^\nu} c_i^2 + \sum_{n=2}^{\nu} n^2 \sum_{i=2^{n-1}+1}^{2^n} c_i^2 \right)$$

$$\le K_4 \left(c_1^2 + \sum_{n=1}^{2^\nu} c_n^2 \ln^2 n \right),$$

whence (10) follows.

In the case $|c_1| \ge |c_2| \ge \dots$ there is a parallel lower bound:

$$\|\{c_n\}\| \ge K_5 \left(c_1^2 + \sum_{n>1} c_n^2 \ln^2 n \right)^{1/2}, \quad K_5 > 0. \tag{11}$$

Indeed, it follows from Menshov's Lemma 2 that

$$I(\underbrace{1, \dots, 1}_{n}) \ge K_6 \, n \ln^2 n.$$

Therefore, considering properties (iv) and (v), we have

$$I(c_1, \dots, c_{2^\nu}) \ge I(c_1) + I(c_2) + I(c_3, c_4) + \dots + I(c_{2^{\nu-1}+1}, \dots, c_{2^\nu})$$

$$\ge c_1^2 I(1) + c_2^2 I(1) + c_4^2 I(1, 1) + \dots + c_{2^\nu}^2 I(1, \dots, 1)$$

$$\ge K_7(c_1^2 + c_2^2) + K_6 \sum_{q=1}^{\nu-1} 2^q q^2 c_{2^q+1} \ge K_5 \left(c_1^2 + \sum_{n>1} c_n^2 \ln^2 n \right).$$

The class of monotone sequences in $\mathfrak{S}_\Omega$ is thus described exhaustively.

Theorem 2 [142]. *Let a sequence of coefficients $\{c_n\}$, $|c_n| \downarrow$, be given. Then for series (1) to converge almost everywhere for every ONS ϕ, the condition $\sum c_n^2 \ln^2 n < \infty$ is necessary and sufficient.*

This fact follows immediately from Theorem 1 and inequalities (10) and (11).

For nonmonotone sequences, inequality (11) is not generally true. Some refinements of estimate (10) are known (see [144, 37]), but a conclusive result as powerful as Theorem 2 has not been found.

Analogous questions have been considered in the class $\mathfrak{B}$ of uniformly bounded ONS. As is known (Menshov), this class also contains a system for which $\{\ln^2 n\}$ is an exact Weyl multiplier.

Tandori investigated in detail the class of sequences $\mathfrak{S}_\mathfrak{B} = \bigcap_{\phi \in \mathfrak{B}} \mathfrak{S}(\phi)$ and established for it the analogues of the foregoing results. In particular, it was proved [142] that Theorem 2 remains valid for this class of systems. The same also holds for the class of systems orthonormal on $[0,1]$ and bounded by a constant $M > 1$. The case $M = 1$ presents special difficulties and has not yet been solved.[*]

We should note that in these problems the additional condition that the system ϕ be complete and that the functions it comprises be smooth does not change the situation. D. E. Menshov [75] constructed a uniformly bounded ONS consisting of algebraic polynomials and having $\{\ln^2 n\}$ as an exact Weyl multiplier. Menshov's method was further developed by L. Leindler [67] for extending theorems on divergence of orthogonal series to polynomial systems.

Facts about the stability of the orthogonalization operator (Chap. I, § 5) are useful in problems of this type. Indeed, they imply the possibility of approximating any ONS $\{\phi_n\}$ by a system of algebraic polynomials $\{\psi_n\}$, also orthonormal. so closely that $\sum \|\phi_n - \psi_n\|_2^2 < \infty$. This condition obviously ensures the equiconvergence almost everywhere of the series $\sum c_n \phi_n$ and $\sum c_n \psi_n$ if $\sum c_n^2 < \infty$. Given an ONS with exact Weyl multiplier $\omega(n)$, this approach allows one to construct automatically a complete polynomial ONS with the same exact Weyl multiplier.

Unconditional convergence. A series of functions is said to be *unconditionally convergent almost everywhere* if it converges almost everywhere for every ordering of the terms. The exceptional set of measure zero where divergence is allowed may depend on the order of the terms.

Coefficient conditions sufficient for the unconditional convergence of orthogonal series were investigated by Menshov and Orlicz (see [55]).

Exhaustive results in this area were obtained by Tandori [141]. He proved the following theorem.

Theorem 3. *Given a sequence of coefficients $\{c_n\}$, series (1) is unconditionally convergent almost everywhere for all ONS if and only if the coefficients satisfy the condition*

$$\sum_k \left[\sum_{n=2^{2^k}+1}^{2^{2^{k+1}}} (c_n^*)^2 \ln^2 n \right]^{1/2} < \infty, \tag{12}$$

where c_n^* is the nonincreasing rearrangement of the numbers $|c_n|$.

This in particular implies the impossibility of improving Orlicz's conditions (see [55]), which suffice for a sequence $\omega(n)$ to be a Weyl multiplier for unconditional convergence. In essence Theorem 3 describes the class of sequences belonging to $\mathfrak{S}_\Omega$ under every rearrangement.

---

[*] Recently B. Kashin showed that in this case $\omega(n) = \ln^2 n$ remains an exact Weyl multiplier.

The sufficiency of condition (12) follows easily from Lemma 1; the main burden is to prove necessity. The theorem formulated above remains valid in the class of uniformly bounded ONS (see [147]).

§ 2. Garsia's Theorem

The existence of examples of divergent Fourier series in L^2 naturally leads to the question of whether such examples are typical or exceptional. What are the facts for "most" series

$$\sum c_n \phi_n(x), \quad \sum c_n^2 < \infty, \tag{1}$$

in orthonormal systems?

There are various ways of making this question concrete. The classical formulation is as follows: for an arbitrary series (1) in an ONS, we examine the random series

$$\sum \pm c_n \phi_n(x) \tag{2}$$

with the $+$ and $-$ signs being chosen independently and with equal probability. The question is, with what probability does series (2) converge almost everywhere? The answer to this question comes from the properties of the Rademacher system, which lets us write series (2) in the form

$$\sum c_n r_n(t) \phi_n(x). \tag{3}$$

The probability space here has the natural realization as the set I of dyadic-irrational points t of the interval $[0,1]$ with Lebesgue measure.

The condition $\{c_n\} \in l_2$ implies that $\sum c_n^2 \phi_n^2(x) < \infty$ almost everywhere. Therefore for almost all x series (3) converges almost everywhere. Using Fubini's theorem, we conclude that for almost all t the series converges almost everywhere in x; that is, we obtain the following result, due to Zygmund and Paley:

For any ONS ϕ and any sequence $\{c_n\} \in l_2$, series (2) converges almost everywhere with probability 1. If the system is uniformly bounded, the sum of the series almost certainly belongs to $\bigcap_{p < \infty} L^p$.

The latter assertion follows from the Khinchin inequality

$$\int_0^1 |\sum c_n r_n(t) \phi_n(x)|^p dt \leq C_p (\sum c_n^2 \phi_n^2(x))^{\frac{p}{2}}$$

after integration with respect to x and application of Fubini's theorem. Moreover, if $\sup_n \|\phi_n\|_p < \infty \, (2 \leq p < \infty)$, *then the condition $\sum c_n^2 < \infty$ ensures the convergence of series (2) in the L^p-metric with probability 1* (Orlicz).

The proof is analogous to the preceding one; it suffices to use the inequality

$$\int_0^1 \sup_{1 \leq l < \infty} \left| \sum_{i=1}^l d_i r_i(t) \right|^p dt \leq C_p (\sum d_i^2)^{\frac{p}{2}}$$

which follows, for example, from inequality (8) of Chap. III, § 3, for the Haar system.

These are elementary facts of the theory of random Fourier series (see [57]). The methods of this theory are used in particular for constructing "typical" counterexamples. We shall encounter a result of this type in Chap. III (§ 4, Theorem 6).

It was recently discovered that the above result on convergence almost everywhere is again obtained if instead of random distribution of signs in series (1) we consider random rearrangement of the series. Namely, the following theorem, established by Garsia [40], is true.

Theorem 1. *For any given series* (1) *in an ONS, there is some rearrangement of the terms that makes the series converge almost everywhere. Moreover, the convergence occurs for most rearrangements.*

The meaning of the last assertion will be made more precise in what follows.

A beautiful combinatorial lemma due to Spitzer [125] plays a central role in the proof. In order to state it, we introduce the following notation.

For every number a define $a^+ = \max\{0, a\}$. P_n is the set of all permutations σ of the elements $(1, 2, \ldots, n)$. If $\mathbf{x} = (x_1, \ldots, x_n) \in \mathbb{R}^n$ is an arbitrary vector and $\sigma = (\sigma_1, \ldots, \sigma_n) \in P_n$, then we shall denote by $\sigma \mathbf{x}$ the vector $(x_{\sigma_1}, x_{\sigma_2}, \ldots, x_{\sigma_n})$.

It is easily seen that every permutation σ factors uniquely as a product of cycles (cyclic permutations): $\sigma = \alpha_1 \ldots \alpha_l$. For example, $(1567432) = (1)(2547)(36)$. Set

$$\delta(\sigma, \mathbf{x}) = \max_{1 \leq k \leq n} \left[\sum_{i=1}^{k} x_{\sigma_i} \right]^+ ; \quad \gamma(\sigma, \mathbf{x}) = \sum_{\nu} \left[\sum_{i \in \alpha_\nu(\sigma)} x_i \right]^+ . \tag{4}$$

Lemma 1 [125]. *Given an arbitrary vector* $\mathbf{x}$, *there exists a one-to-one mapping* $T: P_n \to P_n$ *such that*

$$\delta(\sigma, \mathbf{x}) = \gamma(T\sigma, \mathbf{x}). \tag{5}$$

The functions δ and γ are continuous in $\mathbf{x}$, so it is enough to examine an everywhere dense set of vectors $\mathbf{x}$; for example, those whose coordinates are linearly independent over the rational numbers. Fix such a vector $\mathbf{x}$ and take an arbitrary permutation σ. Consider the broken line in $\mathbb{R}^2$ with corners at $(0,0), (1, x_{\sigma_1}), (2, x_{\sigma_1} + x_{\sigma_2}), \ldots, (n, x_{\sigma_1} + x_{\sigma_2} + \cdots + x_{\sigma_n})$. Construct its minimal convex majorant, and let $0 = s_0 < \cdots < s_l = n$ be the abscissas of the corners of the resulting broken line. Then the following relations are evident:

$$\max_{s_{\nu-1} < s \leq s_\nu} \frac{1}{s - s_{\nu-1}} \sum_{i = s_{\nu-1}+1}^{s} x_{\sigma_i} = \frac{1}{s_\nu - s_{\nu-1}} \sum_{i = s_{\nu-1}+1}^{s_\nu} x_{\sigma_i} \equiv \lambda_\nu \quad (1 \leq \nu \leq l);$$

$$\lambda_\nu > \lambda_{\nu+1} \quad (1 \leq \nu < l). \tag{6}$$

Geometric considerations make it clear (in view of the linear independence over the rationals) that the indices s_ν are uniquely determined by relations (6). The permutation σ will correspond to the new permutation $\tau = T\sigma$, equal to the product of the cycles $\alpha_\nu = (\sigma_{s_{\nu-1}+1}, \ldots, \sigma_{s_\nu})$. The definitions (4) and the relations (6) immediately give equation (5). To prove that this mapping is one-to-one, it suffices to prove surjectivity: $T(P_n) = P_n$.

Fix a permutation $\tau \in P_n$ and decompose it into cycles $\tau = \alpha_1 \ldots \alpha_l$, ordering the factors α_ν so that the averages over the cycles $\dfrac{1}{r_\nu} \sum_{i \in \alpha_\nu} x_i$ (r_ν being the number of elements in α_ν) are strictly increasing. The initial element τ_1^ν in the notation of each cycle $\alpha_\nu = (\tau_1^\nu \ldots \tau_{r_\nu}^\nu)$ can be chosen so that the following equations hold:

$$\max_{1 \le k \le r_\nu} \frac{1}{k} \sum_{i=1}^{k} x_{\tau_i^\nu} = \frac{1}{r_\nu} \sum_{i=1}^{r_\nu} x_{\tau_i^\nu}. \tag{7}$$

This comes from the following proposition: given numbers $y_1, \ldots, y_r$, there is some index $q \in [1, r]$ making the inequalities

$$\frac{1}{k} \sum_{i=1}^{k} y_{q+i} \le \frac{1}{r} \sum_{i=1}^{r} y_i \quad (1 \le k \le r)$$

true. (The indices are added modulo r.)

For proof of this last proposition, observe that without loss of generality we can assume the right-hand side to be zero. In this case it is manifestly sufficient to pick an index q according to the condition

$$\sum_{i=1}^{q} y_i = \max_{1 \le m \le r} \sum_{i=1}^{m} y_i.$$

Define the permutation $\sigma = (\tau_1^1 \ldots \tau_{r_1}^1 \tau_1^2 \ldots \tau_{r_2}^2 \ldots \tau_1^l \ldots \tau_{r_l}^l) \in P_n$. It follows immediately from the construction that $T\sigma = \tau$. Lemma 1 is proved.

Lemma 2 [40]. *Let $\{\phi_i\}$ ($1 \le i \le n$) be an ONS on X, and let $\{c_i\}$ be any given numbers. Then we have the inequality*

$$\frac{1}{n!} \sum_{\sigma \in P_n} \int_X S^*(\sigma, t) dt \le 4 \sqrt{\mu X} \left(\sum_{1}^{n} c_i^2 \right)^{1/2}, \tag{8}$$

where

$$S^*(\sigma, t) = \max_{1 \le k \le n} \left| \sum_{i=1}^{k} c_{\sigma_i} \phi_{\sigma_i}(t) \right|.$$

Define

$$S_+(\sigma, t) = \max_{1 \le k \le n} \left[\sum_{i=1}^{k} c_{\sigma_i} \phi_{\sigma_i}(t) \right]^+, \quad S_-(\sigma, t) = \max_{1 \le k \le n} \left[- \sum_{i=1}^{k} c_{\sigma_i} \phi_{\sigma_i}(t) \right]^+.$$

Let $\Theta_\alpha(\tau) = 1$ if the cycle α is a factor of the permutation τ, and $\Theta_\alpha(\tau) = 0$ otherwise. Application of Lemma 1 together with the obvious inequality $S^* \le S_+ + S_-$ shows that

$$\sum_{\sigma \in P_n} S^*(\sigma, t) \le \sum_{\tau \in P_n} \sum_{\alpha} \left[\sum_{i \in \alpha} c_i \phi_i(t) \right]^+ \Theta_\alpha(\tau) + \sum_{\tau} \sum_{\alpha} \left[- \sum_{i \in \alpha} c_i \phi_i(t) \right]^+ \Theta_\alpha(\tau)$$

$$\le 2 \sum_{\alpha} \left| \sum_{i \in \alpha} c_i \phi_i(t) \right| \sum_{\tau} \Theta_\alpha(\tau). \tag{9}$$

Note that each cycle α containing r elements ($|\alpha|=r$) occurs in exactly $(n-r)!$ distinct permutations τ; that is $\sum_\tau \Theta_\alpha(\tau)=(n-r)!$. We therefore obtain from (9) upon integration

$$\sum_\sigma \int_X S^*(\sigma,t)\,dt \le 2 \sum_{r=1}^n (n-r)! \sum_{\alpha;\,|\alpha|=r} \int_X \left|\sum_{i\in\alpha} c_i \phi_i(t)\right| dt.$$

Considering, further, that each choice of an index sequence $1\le i_1 < i_2 < \cdots < i_r \le n$) generates $\dfrac{r!}{r}$ cycles, we see that

$$\frac{1}{n!}\sum_\sigma \int_X S^*(\sigma,t)\,dt \le 2 \sum_{r=1}^n \frac{1}{r\binom{n}{r}} \sum_{1\le i_1<\cdots<i_r\le n} \int_X \left|\sum_{k=1}^r c_{i_k}\phi_{i_k}(t)\right| dt$$

$$\le 2\sqrt{\mu X}\sum_{r=1}^n \frac{1}{r}\binom{n}{r}^{-1} \sum_{1\le i_1<\cdots<i_r\le n}\left(\sum_{k=1}^r c_{i_k}^2\right)^{1/2}$$

$$\le 2\sqrt{\mu X}\sum_{r=1}^n \frac{1}{r}\left[\binom{n}{r}^{-1} \sum_{1\le i_1<\cdots<i_r\le n}\left(\sum_{k=1}^r c_{i_k}^2\right)\right]^{1/2}.$$

(The Bunyakovskiĭ-Cauchy inequality is applied first to the integrals and then to the sum.) Finally, using the identity

$$\sum_{1\le i_1<\cdots<i_r\le n}\left(\sum_{k=1}^r c_{i_k}^2\right) = \binom{n-1}{r-1}\sum_{i=1}^n c_i^2,$$

we conclude that

$$\frac{1}{n!}\sum_\sigma \int_X S^*(\sigma,t)\,dt \le 2\sqrt{\mu X}\,n^{-1/2}\sum_{r=1}^n r^{-1/2}\left(\sum_1^n c_i^2\right)^{1/2}$$

from which (8) follows.

Now it is not hard to prove that series (1) converges almost everywhere after some rearrangement of terms, and that this is true for "most" permutations. Let us make the latter assertion more precise. Fix a sequence of index numbers $\{v_l\}$ such that

$$\sum_l \left(\sum_{i=v_{l-1}+1}^{v_l} c_i^2\right)^{1/2} < \infty. \tag{10}$$

Consider the set $P_{\{v_l\}}$ of all permutations $\sigma=\{\sigma_1,\ldots\}$ of the natural numbers satisfying the condition $v_{l-1}<\sigma_i\le v_l$ whenever $v_{l-1}<i\le v_l$. Each element $\tau\in P_{\{v_l\}}$ is given by a sequence $\{\sigma^{(l)}\}$ ($l=1,2,\ldots$), where $\sigma^{(l)}$ is an arbitrary permutation of the elements in the l-th block $(v_{l-1}+1,\ldots,v_l)$. Thus, the set $P_{\{v_l\}}$ is is naturally identified with the direct product of the sets P_{n_l} ($n_l=v_l-v_{l-1}$). Introducing the measure μ_n on P_n in the usual manner (that is, setting $\mu_n E=\dfrac{m}{n!}$ for

any set $E \subset P_n$ consisting of m elements), we make $P_{\{v_l\}}$ into a probability space with the measure $\hat{\mu} = \oplus \mu_{n_l}$ (see [50]).

The exact statement of Theorem 1 is thus:

for almost all permutations $\sigma \in P_{\{v_l\}}$, *series* (1) *converges almost everywhere.*

Indeed, let

$$S_v(\sigma, t) = \sum_{i=1}^{v} c_{\sigma_i} \phi_{\sigma_i}(t); \quad S_l^*(\sigma, t) = \max_{v_{l-1} < k \le v_l} \left| \sum_{i=v_{l-1}+1}^{k} c_{\sigma_i} \phi_{\sigma_i}(t) \right|.$$

For any permutation σ belonging to $P_{\{v_l\}}$, the definition of the set $P_{\{v_l\}}$ implies that $S_{v_l}(\sigma, t) - S_{v_{l-1}}(\sigma, t) = \sum_{i=v_{l-1}+1}^{v_l} c_i \phi_i(t)$, whence, by virtue of (10), $\sum \|S_{v_l} - S_{v_{l-1}}\|_2 < \infty$; so the partial sums $S_{v_l}(\sigma, t)$ converge almost everywhere with respect to t. At the same time, on the basis of Lemma 2 we have

$$\sum_l \int_{P_{\{v_l\}}} \left[\int_X S_l^*(\sigma, t) dt \right] d\hat{\mu} = \sum_l \frac{1}{n_l!} \sum_{\sigma^{(l)}} \int_X S_l^*(\sigma^{(l)}, t) dt$$

$$\le 4\sqrt{\mu X} \sum_l \left[\sum_{i=v_{l-1}+1}^{v_l} c_i^2 \right]^{1/2} < \infty.$$

Hence it follows that the series $\sum S_l^*(\sigma, t)$ converges almost everywhere for almost all σ, and therefore $S_l^*(\sigma, t) = o(1)$. The theorem now follows.

Garsia's theorem means that the terms of any orthogonal Fourier series from L^2 can be arranged in such an order as to make the series converge almost everywhere. This is not so for Fourier series from L^p for $p < 2$ [98] (see Chap. IV, § 2).

Note that the method of ordering in Theorem 1 is defined specifically for an individual series (1); it depends in an essential way not only on the system $\{\phi_n\}$ but also on the coefficients $\{c_n\}$. Thus the following problem, which goes back to A. N. Kolmogorov, remains open: can an arbitrary ONS be numbered in such a way as to make it a system of convergence? (For Cesàro summability this has been answered in the affirmative by D. E. Menshov; see [47].)

This question opens up one of the most interesting aspects of the whole set of problems concerning the connection between the ordering of an orthonormal system of functions and the behavior of Fourier series. It should be noted that general systems, in contrast to the classical systems, have no natural numbering.

Another question in this same area, that of the existence of a complete ONS that is a system of convergence independently of the ordering of its elements, was answered in the negative a few years ago [90, 153]. These results will be discussed in § 2 of Chap. III.

In conclusion, we mention a generalization of Theorem 1 to nonorthogonal series. E. M. Nikishin, elaborating Garsia's technique, showed (see [84]) that if a series of functions

$$\sum f_n(t) \tag{11}$$

has a subsequence of partial sums converging almost everywhere and satisfies the condition

$$\sum f_n^2(t) < \infty \quad \text{almost everywhere,} \tag{12}$$

then after some rearrangement it converges almost everywhere. This result has been applied in investigation of the set $\mathscr{I}(f)$ of functions F that are the sums of a given series (11) under all the (convergent almost everywhere) rearrangements of its terms. Specifically, it has been proved [84] that *given condition* (12), *the set* $\mathscr{I}(f)$ *is linear and closed under convergence almost everywhere.*

In addition to this, Nikishin [85] constructed *an example of a series* (11) *satisfying the condition* $\sum |f_n(x)|^{2+\varepsilon} < \infty$ $(\forall \varepsilon > 0)$ *for which the set* $\mathscr{I}(f)$ *is not even convex.* It gives the answer to a problem of Banach.

§ 3. The Coefficients of Convergent Series in Complete Systems

In § 1 we discussed the intersection of the classes $\mathfrak{S}(\phi)$ over the set of all orthonormal (bounded, complete) systems. In this section we shall examine the class

$$\mathfrak{S}^{\Pi} = \bigcup_{\phi \in \Pi} \mathfrak{S}(\phi), \tag{1}$$

where Π is the set of all complete ONS on X, and some other classes of this sort.

The problem is this: suppose it is known that a series

$$\sum c_n \phi_n(x) \tag{2}$$

in a complete ONS converges almost everywhere. What can be said about the coefficients c_n? The formulation of the question is due to W. Orlicz [108], who pointed out the following necessary condition:

$$\liminf \frac{|c_n|}{n} = 0. \tag{3}$$

On the other hand, for the classical complete systems the convergence of series (2) almost everywhere implies the condition

$$\liminf |c_n| = 0. \tag{4}$$

The question arises as to whether this is true in general. In particular, can the coefficients of a convergent series in a complete system grow to ∞?

We note that the necessity of condition (3) is a consequence of the following theorem (Orlicz [108]):

If $\phi \in \Pi$, then

(i) $\sum \phi_n^2(x) = \infty$ almost everywhere,

and (what is stronger) even

(ii) $\sum \left[\int_E |\phi_n| dx \right]^2 = \infty$ for any set $E \subset X, \mu E > 0$.

This theorem permits a strengthening of the necessary condition (3). Interestingly enough, the condition (5) which is thus obtained is definitive. The following proposition is true [92].

Theorem 1. *If series* (2), *where* $\phi \in \Pi$, *converges almost everywhere (or on a set E of positive measure, or in the metric of* $L^p(E)$, $p \geq 1$), *then the coefficients satisfy the condition*

$$\sum \frac{1}{c_n^2 + 1} = \infty. \tag{5}$$

Conversely, for any sequence $\{c_n\}$ *satisfying condition* (5), *it is possible to exhibit an ONS* $\{\phi_n\}$, *complete in* L, *such that series* (2) *converges almost everywhere (and in the* L^p-metric for every $p \in [1,2)$).

The first assertion of the theorem follows easily from (i) and (ii). The proof of the converse proposition rests on the following lemma [92].

Lemma. *Suppose we are given pairwise disjoint sets* X_j $(j = 1, 2, \ldots)$, $\mu X_j > 0$, $\bigcup_j X_j = X$, *and on each of them there is defined an ONS* $\psi^j = \{\psi_i^j\}$, *complete in* $L^p(X_j)$, $p \geq 1$. *Suppose further that there is given a sequence of orthogonal matrices* $A_k = \|a_{ij}^{(k)}\|$, $1 \leq i, j \leq l_k$, $l_k \uparrow \infty$. *Define*

$$\phi_i^{(k)}(x) = \begin{cases} a_{ij}^{(k)} \psi_{k-m(j)}^{(j)}(x), & x \in X_j, \quad 1 \leq j \leq l_k; \\ 0, & x \in X_j, \quad j > l_k \end{cases} (1 \leq i \leq l_k), \tag{6}$$

where the numbers $m(j)$ *are determined by the inequality* $l_{m(j)} < j \leq l_{m(j)+1}$, *and* $l_0 = 0$. *Then the functions* $\{\phi_i^{(k)}\}$, $1 \leq i \leq l_k$, $k = 1, 2, \ldots$, *form an ONS complete in* $L^p(X)$.

That the system ϕ is orthonormal is verified immediately. Next, if $f \in L^p$ and $\int_X f \phi_i^{(k)} dx = 0$ ($\forall i$), then by virtue of the nonsingularity of the matrices A_k we have that $\int_{X_j} f \psi_{k-m(j)}^{(j)} dx = 0$ ($1 \leq j \leq l_k$). These relations, which hold for every k, show that on each X_j the function f is orthogonal to all the elements of the system ψ^j, so in fact $f = 0$.

Now let a sequence $\{c_n\}$ satisfying condition (5) be given. We shall construct a complete ONS ϕ such that series (2) converges almost everywhere. We can, of course, assume $c_n \neq 0$ (this is achieved by small perturbations of the coefficients, without affecting the convergence). In this case condition (5) can be rewritten in the form $\sum c_n^{-2} = \infty$. We can define inductively a sequence of index numbers $\{v_k^{(j)}\}$ ($0 \leq j \leq k$) satisfying the conditions

$$v_{k-1}^{(k-1)} = v_k^{(0)} < v_k^{(1)} < \cdots < v_k^{(k)} \quad (\forall k); \quad v_1^{(0)} = 0;$$

$$\lambda_k^{(j)} = \left[\sum_{n=v_k^{(j-1)}+1}^{v_k^{(j)}} c_n^{-2} \right]^{1/2} > k^2 \quad (1 \leq j \leq k; \, k = 1, 2, \ldots). \tag{7}$$

For each k define an orthogonal matrix A_k of order $l_k = v_k^{(k)} - v_k^{(0)} \geq k$. Set

$$a_{ij}^{(k)} = \begin{cases} \dfrac{(-1)^i}{\lambda_k^{(j)}} \dfrac{1}{c_k}, & i = n - v_k^{(0)}, \ v_k^{(j-1)} < n \leq v_k^{(j)} \quad (1 \leq j \leq k) \\[2mm] 0 & \text{for all other } i. \end{cases} \tag{8}$$

This defines the first k columns of the matrix; the remaining columns are chosen arbitrarily, subject only to the requirement of orthogonality.

Next, choose a sequence $\{\gamma_j\}$, $1 = \gamma_0 > \gamma_1 > \ldots \to 0$, and on each of the intervals $X_j = (\gamma_j, \gamma_{j-1}]$ consider the ONS ψ^j, complete in L, that one obtains from the trigonometric system by a linear change of variable and normalization. Finally, we define the system $\{\phi_i^{(k)}\}$ in accordance with (6). Letting $\phi_n = \phi_i^{(k)}$, $n = v_k^{(0)} + i$, $1 \leq i \leq l_k$, we obtain the ONS $\{\phi_n\}$, which is complete in $L[0,1]$.

Let us show that series (2) converges uniformly on each X_j. For this it is enough, having fixed j and $k \geq j$, to note the following relations, which follow from (6) and (8):

$$\max_{v_k^{(j-1)} < v \leq v_k^{(j)}} \left\| \sum_{n = v_k^{(j-1)} + 1}^{v} c_n \phi_n(x) \right\|_{L^\infty(X_j)} = \frac{1}{\lambda_k^{(j)}} \| \psi_{k-m(j)}^{(j)} \|_{L^\infty(X_j)},$$

$$c_n \phi_n(x) = 0, \quad n \in [v_k^{(0)}, v_k^{(k)}] \setminus [v_k^{(j-1)} + 1, v_k^{(j)}], \quad x \in X_j;$$

and then to recall the uniform boundedness of the system ψ^j, and inequality (7). It is not difficult to show (see [92]) that if the numbers γ_j decrease sufficiently rapidly (the choice of these numbers being made after the other parameters of the construction are already fixed), then the series also converges in L^p, $1 \leq p < 2$.

Remarks. 1. It follows from Theorem 1 that the functions $\{\phi_n\}$ forming a complete ONS can tend to zero as $n \to \infty$, and can do so as rapidly as allowed by Orlicz's conditions (i) and (ii). More precisely: *for any sequence of $d_n > 0$ such that $\sum d_n^2 = \infty$, it is possible to construct a complete ONS $\{\phi_n\}$ satisfying the conditions*

$$|\phi_n(x)| = o_x(d_n) \ (\forall x), \qquad \|\phi_n\|_p = o(d_n) \ (\forall p < 2).$$

To prove this, it suffices to let $c_n = 1/d_n$ and to apply the preceding theorem. The impossibility of improving relation (i) can be illustrated further by the following example (see [92]), which is also obtained with the aid of the construction of Lemma 1: *there exists a complete ONS $\{\phi_n\}$ for which the series*

$$\sum_n |\phi_n(x)|^p \operatorname{sgn} \phi_n(x)$$

converges everywhere for all $p > 0$.

In addition, the relation mentioned can be made more exact, as follows (V. Ya. Kozlov; see [47]): *if $\phi \in \Pi$, then $\sum [\phi_n^+(x)]^2 = \infty$ and $\sum [\phi_n^-(x)]^2 = \infty$ almost everywhere* (here $a^+ = \max(a,0)$, $a^- = \min(a,0)$).

2. If series (2) of a complete ONS *unconditionally* is convergent almost everywhere (or at least in measure), then the coefficients cannot tend to ∞. In this case condition (4) holds (P. L. Ulyanov [161]). Indeed, as Orlicz showed [109], unconditional convergence of the series $\sum f_n(x)$ in measure implies the finiteness of $\sum f_n^2(x)$ almost everywhere, after which one has only to use property (i).

3. The system ϕ in Theorem 1 depends on the particular sequence of coefficients $\{c_n\}$. This is unavoidable inasmuch as the inclusion $l_0 \subset \mathfrak{S}(\phi)$ does not hold for any complete ONS (Orlicz; see [55]) ($l_0 = \{c; c_k = o(1)\}$), nor even the weaker inclusion $l_p \subset \mathfrak{S}(\phi)$ (for $p > 2$) (see [94]).

Nevertheless, if certain regularity conditions are imposed on the numbers $\{c_n\}$, then the situation changes. For example,

for any sequence of $d_n > 0$ such that $\sum d_n^{-2} = \infty$, it is possible to construct a complete ONS $\{\phi_n\}$ such that the class $\mathfrak{S}(\phi)$ contains all convex sequences $\{c_n\}$, $c_n = O(d_n)$ (see [92]).

We shall further say that a sequence $\{c_n\}$ belongs to the class $\mathfrak{S}_*(\phi)$, $\mathfrak{S}_p(\phi)$, $\mathfrak{S}_{mes}(\phi)$ or $\mathfrak{F}_p(\phi)$ if series (2) for the given system ϕ respectively converges on a set of positive measure, in the L^p-metric, in measure, or is the Fourier series of a function in L^p. Moreover, if $\mathfrak{A}(\phi)$ is any of these classes and Φ is some set of ON systems, we shall write $\mathfrak{A}^\Phi = \bigcup_{\phi \in \Phi} \mathfrak{A}(\phi)$.

Theorem 1 gives the following equality:

$$\mathfrak{S}^\Pi = \mathfrak{S}_*^\Pi = \mathfrak{S}_p^\Pi = \left\{ c; \sum \frac{1}{c_n^2 + 1} = \infty \right\} \quad (p < 2).$$

For $p \geq 2$, as is known, $\mathfrak{F}_p^\Pi = \mathfrak{S}_p^\Pi = l_2$ (use the Paley-Zygmund theorem; see § 2). At the same time, the class $\mathfrak{S}_{mes}^\Pi$ contains all sequences of numbers.

Theorem 2. *For any sequence $\{c_n\}$, it is possible to exhibit an ONS ϕ, complete in L, such that series (2) converges in measure to some function $f \in \bigcap_{p < 2} L^p$ and is the Fourier series of this function.*

The proof (see [92]) is based on the construction of Lemma 1. The statement given implies also that there are no restrictions on the Fourier coefficients (with respect to complete ONS) of functions from the classes L^p, $p < 2$.

A few words about results relating to the class $\mathfrak{B}$ of bounded ONS and the class $\Pi\mathfrak{B}$ of complete bounded ONS.

As is known, $\mathfrak{S}_{mes}^\mathfrak{B} \subset l_0$ (see [55]). It is not doubted that in fact $\mathfrak{S}^{\Pi\mathfrak{B}} = l_0$, although this, it seems, has not been proved. We do know the characterization

$$\mathfrak{S}_*^{\Pi\mathfrak{B}} = \left\{ c; \sum \frac{1}{c_n^2 + 1} = \infty \right\}.$$

The main point in the proof (see [103]) is the establishment of the fact that the system $\phi = \phi(\{c_n\}; x)$ constructed in Theorem 1 can be extended from the interval $X_1 = (1/2, 1]$ to a complete bounded ONS on $X = [0, 1]$.

Calculations show the classes $\mathfrak{F}_p^\mathfrak{B}$, $\mathfrak{F}_p^{\Pi\mathfrak{B}}$, $1 < p < 2$, to be as follows:

$$\mathfrak{F}_p^\mathfrak{B} = \mathfrak{F}_p^{\Pi\mathfrak{B}} = \mathfrak{P}_p \equiv \{ c; \sum (c_n^*)^p n^{p-2} < \infty \}$$

(c_n^* being the nonincreasing rearrangement of the numbers $|c_n|$). The containment $\mathfrak{F}_p^\mathfrak{B} \subset \mathfrak{P}_p$ is Paley's theorem (see [55]); the converse containment $\mathfrak{P}_p \subset \mathfrak{F}_p^{\Pi\mathfrak{B}}$ follows from a theorem of Hardy and Littlewood (see Chap. I, § 3). which gives the stronger result that $\mathfrak{P}_p \subset \mathfrak{F}_p^\mathfrak{T}$, where $\mathfrak{T}$ is the set of all rearrangements of the trigonometric system. For $p = 1$ we have that $\mathfrak{F}_1^{\Pi\mathfrak{B}} = l_0$.

(This is easily deduced from the lemma of § 3, Chap. I.) This result, however, is not true for the class $\mathfrak{T}$: the containment $\widetilde{\mathfrak{F}}_1^{\mathfrak{T}} \subset l_0$ is a proper containment (Kahane [58]).

If $c_n \downarrow$, then the conditions $c \in \mathfrak{S}_p^{\Pi\mathfrak{B}}$ and $c \in \widetilde{\mathfrak{F}}_p^{\Pi\mathfrak{B}}$ $(1 < p < 2)$ are equivalent, as follows from the same theorem of Hardy and Littlewood. It turns out that this is not the case for $p = 1$. Indeed, as shown in § 3, Chap. I, if $c_n \downarrow$, then

$$c \in \mathfrak{S}_1^{\Pi\mathfrak{B}} \text{ if and only if } c_n = o\left(\frac{1}{\ln n}\right).$$

For $p \geq 2$, as before, $\widetilde{\mathfrak{F}}_p^{\Pi\mathfrak{B}} = \mathfrak{S}_p^{\Pi\mathfrak{B}} = l_2$.

§ 4. Extension of a System of Functions to an ONS

The questions examined in the preceding section are connected with the problem of extending a sequence of functions defined on some set $E \subset X$ to an ON system (complete, bounded) on X. Problems of this sort arise not infrequently in the theory of orthogonal and nonorthogonal expansions (see, for example, [55, 66, 83, 54]; see also Chap. IV, § 2).

The classical result on this topic is due to I. Schur (see [55]).

Theorem 1. *Suppose there is given a sequence of functions $\{f_n\}$ on a set $E \subset X$, $\mu \mathbin{[} E > 0$. These functions can be extended to an ONS on X if and only if the following condition is fulfilled:*

$$\left| \sum_{n=1}^{s} \sum_{k=1}^{s} g_{nk} \xi_n \xi_k \right| \leq \sum_{k=1}^{s} \xi_k^2 \quad (\forall s, \xi_1, \ldots, \xi_s). \tag{1}$$

Here $G = \|g_{nk}\|$ is the Gram matrix of the system f; i.e., $g_{nk} = \int_E f_n f_k \, dx$.

The necessity of condition (1) is obvious: if the system f is extended to an ONS ϕ, then the left side of inequality (1) is equal to $\int_E \left| \sum_{k=1}^{s} \xi_k \phi_k(x) \right|^2 dx$.

To prove the converse assertion, observe that condition (1) implies that the matrix $\tilde{G}_s = I - G_s$, $G_s = \|g_{nk}\|$ $(1 \leq n, k \leq s)$ is nonnegative definite (for any s). Therefore in the space $\mathbb{R}^s$ there exist vectors $\mathbf{z}_1, \ldots, \mathbf{z}_s$ with Gram matrix $\tilde{G}_s$ (the rows of the matrix $\tilde{G}_s^{1/2}$ will do). From considerations of isometry it is clear that if the vectors $\mathbf{z}_1, \ldots, \mathbf{z}_{s-1}$ with Gram matrix $\tilde{G}_{s-1}$ are already chosen, then a vector $\mathbf{z}_s$ can be added to them so that the resulting system will have Gram matrix $\tilde{G}_s$. This permits the inductive construction of a sequence of elements $\{\tilde{f}_n\}$ in the Hilbert space $L^2(\mathbin{[} E)$ with Gram matrix $\tilde{G} = I - G$. Setting $\phi_n(x) = f_n(x)$ if $x \in E$ and $\phi_n(x) = \tilde{f}_n(x)$ if $x \in \mathbin{[} E$, we obtain the required extension of the system f.

Consider now the question of whether the system f can be extended to a complete ONS. It is obviously necessary to require completeness of the system f on E. It turns out that further necessary and sufficient conditions for such extendibility are just as simple to formulate in terms of the matrix G.

Theorem 2 [103]. *In order for a given sequence of functions* $f_n \in L^2(E)$, $E \subset X$, *to be extendable to a complete ONS in* $L^2(X)$, *the following conditions are necessary and sufficient:*

  (i) *the system* $\{f_n\}$ *is complete in* $L^2(E)$;
  (ii) $G^2 = G$;
  (iii) *the matrix* $\tilde{G} = I - G$ *has infinite rank.*

Let $\{\phi_n\}$ be a complete ONS on X and let $\phi_n \equiv f_n$ $(x \in E)$. Then condition (i) is obvious. Next, writing Parseval's identity for the functions $\tilde{\Phi}_n = \phi_n \chi_{\complement E}$, where χ_U is the characteristic function of the set U, we have that

$$\tilde{g}_{nk} = \int_{\complement E} \phi_n \phi_k dx = \int_X \tilde{\Phi}_n \tilde{\Phi}_k dx = \sum_s \int_X \tilde{\Phi}_n \phi_s dx \int_X \tilde{\Phi}_k \phi_s dx = \sum_s \tilde{g}_{ns} \tilde{g}_{sk},$$

so (ii) is fulfilled. The preceding equality also shows that the system of vectors $\tilde{g}_n \in l_2$, $\tilde{g}_n = \{\tilde{g}_{nk}\}$ has the same Gram matrix as the system $\phi_n \in L^2(\complement E)$. Since the latter is complete, it follows that the subspace of l_2 generated by the vectors $\tilde{g}_n$ is infinite-dimensional, so (iii) is satisfied.

Now suppose we are given a system of functions $\{f_n(x)\}$, $x \in E$ satisfying all three of these conditions. Choose an arbitrary complete ONS $\{\psi_n\}$ in $L^2(\complement E)$ and define

$$p_n = \sum_{k=1}^{\infty} \tilde{g}_{nk} \psi_k$$

(taking the sum in the L^2-sense). Then, because of (ii), we have

$$\int_{\complement E} p_n p_k dx = \sum_s \tilde{g}_{ns} \tilde{g}_{ks} = \tilde{g}_{nk}.$$

Let P be the subspace of $L^2(\complement E)$ generated by the vectors $\{p_n\}$ (infinite-dimensional, because of (iii)). Let $U : P \to L^2(\complement E)$ be an isometry onto the whole space. Define

$$\phi_n(x) = \begin{cases} f_n(x), & x \in E \\ (U p_n)(x), & x \in \complement E. \end{cases}$$

Then we have

$$\int_X \phi_n \phi_k dx = \int_E f_n f_k dx + \int_{\complement E} U p_n U p_k dx = g_{nk} + \int_{\complement E} p_n p_k dx = \delta_{nk}.$$

To prove that the system ϕ is complete, it suffices to verify Parseval's identity for any complete (not necessarily ON) system. For this purpose let us take the system

$$\{\Phi_n = \phi_n \chi_E, \ \tilde{\Phi}_n = \phi_n \chi_{\complement E}; \ n = 1, 2, \dots\}.$$

It is easy to see that this system is complete on X. We have further that

$$\sum_s \left[\int_X \Phi_n \phi_s dx\right]^2 = \sum_s \left[\int_E \phi_n \phi_s dx\right]^2 = \sum_s g_{ns}^2 = \sum_s g_{ns} g_{sn} = g_{nn} = \int_X \Phi_n^2 dx.$$

An analogous calculation is valid for $\tilde{\Phi}_n$. Thus the system ϕ provides a continuation of the sequence f that meets our requirements.

This theorem is true in the complex case, too. We note that, as can be seen from the proof just given, the hypotheses of the theorem ensure that any continuation to an ONS complete on $\complement E$ is automatically complete on X.

It is interesting to compare the geometric interpretations of Theorems 1 and 2. The first of these theorems shows that a system can be extended to an ONS if and only if its Gram matrix G induces a bounded operator on the Hilbert space with norm ≤ 1. For it to be possible to extend a complete system to a complete ONS, it is necessary and sufficient that this operator be a projection (with infinite co-rank).

We remark in conclusion that it would be interesting to know under what conditions a system can be continued to a complete uniformly bounded ONS.

Chapter III. Properties of Complete Systems; the Role of the Haar System

As we know, the Haar system was the first ONS with respect to which every continuous function has a convergent Fourier expansion.

This system also possesses other remarkable properties. In particular, every integrable function can be expanded into a Fourier-Haar series converging to the function almost everywhere; and if $f \in L^p$, $p \in [1, \infty)$, then the series also converges in the L^p-metric; that is, $\{\chi_n\}$ forms a basis in every L^p-space.

In the last few years it was discovered that the Haar system plays an essential role in the theory of general orthogonal systems.

In this chapter we present a method for investigating arbitrary complete ONS and bases in function spaces, based on some special properties of the Haar system. This method was first used by the author in a concrete situation, in constructing Fourier series that diverge almost everywhere [90, 93] (see § 2), and was later formalized independently of this problem and applied to some other questions [100, 106].

The method referred to here shows that in a number of problems the Haar system has the *best* properties among all complete orthogonal systems. Roughly speaking, *if a Fourier series divergence phenomenon occurs with the Haar system, then such a phenomenon is unavoidable for any complete ONS (or basis).*

The basic idea leading to this conclusion is as follows.

As is known, two functions f and g defined on $X = [0, 1]$ are said to be metrically equivalent if there exists a one-to-one (on a set of full measure) measure-preserving mapping $T: X \to X$ such that $f(Tx) = g(x)$. A mapping T is said to preserve measure if the equalities $\mu E = \mu T(E) = \mu T^{-1}(E)$ hold for every measurable set $E \subset X$.

If the functions f_n and g_n are metrically equivalent for each n and this equivalence is implemented by a single transformation T not depending on n, then it is natural to call the systems $\{f_n\}$ and $\{g_n\}$ isomorphic. Clearly all metric properties of isomorphic systems are identical.

The concept of a weak isomorphism will be important for us.

We shall say that two systems $\{f_n\}$ and $\{g_n\}$ are *weakly isomorphic* if for each n there exists a measure-preserving mapping T_n that is one-to-one on a set of full measure and for which

$$f_k(T_n x) = g_k(x) \quad (1 \le k \le n).$$

Here the metric properties of the systems f and g are no longer necessarily all the same. For example, the completeness of a system is not always preserved by a weak isomorphism. However, it is easy to see that if the series $\sum c_k f_k$

converges in mean or almost everywhere, then the series $\sum c_k g_k$ possesses the same property

It turns out that *for any complete ONS* $\{\phi_n\}$ *it is possible to exhibit a sequence of linear combinations of the form* $p_k = \sum\limits_{i=n_k+1}^{n_{k+1}} \alpha_i \phi_i, \; n_k \uparrow \infty,$ *satisfying the equations*

$$p_k = \tilde{\chi}_k + \tau_k$$

where $\{\tilde{\chi}_k\}$ *is a system weakly isomorphic to the Haar system and* $\|\tau_k\|$ *decreases rapidly.*

Theorems of this sort make it possible in a number of instances to reduce a problem about an arbitrary complete system to the corresponding problem for the concrete system $\{\chi_n\}$, which, by the way, is considerably more accessible to investigation than, for example, the trigonometric system.

We remark that the result just stated reveals a property specific to the Haar system, and is not true of the other classical systems.

§§ 1–4 of this chapter are devoted to the general theory described above and to its applications to various problems.

§ 5 discusses some properties of Haar expansions which are of interest for the general theory of orthogonal systems.

§ 1. The Basic Construction

Suppose we have fixed a strictly increasing sequence of natural numbers $\Lambda = \{\lambda_k\}$ ($\lambda_1 = 0$). We shall associate to it a special system $\chi^\Lambda = \{\chi_k^\Lambda\}$ of Haar polynomials.

The construction will be carried out on the set I of dyadic-irrational points of the interval $[0,1]$. We shall use the following notation: $l(k)$ is the integer such that $2^l < k \leq 2^{l+1}$; $\chi(E; x)$ is the characteristic function of the set E. We shall call intervals of the form $\left(\dfrac{j-1}{2^k}, \dfrac{j}{2^k}\right)$ intervals of rank k. Define

$$\chi_1^\Lambda \equiv 1; \qquad \chi_2^\Lambda = 2^{(-1/2)\lambda_2} \sum_{r=1}^{2^{\lambda_2}} \chi_{\lambda_2}^{(r)}. \tag{1}$$

Suppose the functions χ_s^Λ are defined for $s \leq n$. We shall write

$$E_s^1(\Lambda) = \{x; \chi_s^\Lambda(x) > 0\}; \qquad E_s^2(\Lambda) = \{x; \chi_s^\Lambda(x) < 0\} \quad (s > 1). \tag{2}$$

Define

$$\chi_n^\Lambda(x) = 2^{1/2\,(l(n) - \lambda_n)} \chi(E_{\nu(n)}^{\alpha_n}(\Lambda); x) \sum_{r=1}^{2^{\lambda_n}} \chi_{\lambda_n}^{(r)}(x) \tag{3}$$

where

$$\alpha_n = 1 \quad \text{and} \quad \nu(n) = \frac{n+1}{2} \qquad \text{if } n \text{ is odd};$$

$$\alpha_n = 2 \quad \text{and} \quad \nu(n) = \frac{n}{2} \qquad \text{if } n \text{ is even}.$$

Thus we define by induction the functions χ_n^A and simultaneously the sets $E_n^i(\Lambda)$, $i=1,2$; $n=1,2,\dots$.

These have the following properties:

(i) $\chi_n^A(x) = \begin{cases} (-1)^{i+1} 2^{l(n)/2}, & x \in E_n^i(\Lambda), i=1,2; \\ 0, & x \notin \bigcup_{i=1}^2 E_n^i(\Lambda). \end{cases}$

This fact is evident when one notes that

$$\left| \sum_r \chi_\lambda^{(r)}(x) \right| = 2^{\lambda/2} \quad (x \in I, \lambda = 1,2,\dots). \tag{4}$$

(ii) Each of the sets $E_s^i(\Lambda)$ can be represented as a union of intervals of rank $\lambda_s + 1$.

For $s=2$, this follows from (1). Assuming our assertion for $s<n$, in particular for $s=v(n)$, and using the fact that the set of points where the function $\sum \chi_{\lambda_n}^{(r)}$ does not change sign is a union of intervals of rank $\lambda_n + 1$, we conclude from (3) that property (ii) is true for $s=n$.

Observe that this also implies the following:

$$\chi_n^A = 2^{1/2(l(n)-\lambda_n)} \sum_r \gamma_r \chi_{\lambda_n}^{(r)} \quad (\gamma_r = 0 \text{ or } 1); \tag{5}$$

thus the functions χ_n^A are polynomials in the Haar system.

(iii) $\mu E_s^i = 2^{-l(s)-1}$ $(s=2,3,\dots; i=1,2)$.

For $s=2$ this follows from (1). Assuming this equality true for $s<n$, we have from (3) that

$$\mu(E_n^1 \cup E_n^2) = \mu E_{v(n)}^{\alpha_n} = 2^{-l(v(n))-1}.$$

But the sets E_n^1 and E_n^2 are disjoint and their measures are equal, by (5), so we obtain that

$$\mu E_n^i = \tfrac{1}{2} 2^{-l(v(n))-1} = 2^{-l(n)-1}.$$

We shall call the sets E_n^i, $l(n)=q$, sets of rank q.

The following assertions are deduced by induction from (1)–(3).

(iv) For each q, the sets E_n^i of rank q are disjoint, and their union is the whole set I.

(v) Each of the sets of rank less than q consists of a finite number of sets of rank q.

Using these properties, we shall justify the following proposition.

Lemma. *For any increasing sequence of natural numbers Λ, the corresponding system χ^A is weakly isomorphic to the Haar system.*

We define, by analogy with (2),

$$\delta_n^1 = \{x; \chi_n(x) > 0\}, \qquad \delta_n^2 = \{x; \chi_n(x) < 0\}.$$

Immediately from the definition of the Haar functions we have

(i') $\qquad \chi_n(x) = \begin{cases} (-1)^{i+1} 2^{l(n)/2}, & x \in \delta_n^i \\ 0, & x \notin \bigcup \delta_n^i \end{cases} \quad (n>1).$

(ii') Each of the sets δ_n^i is an interval in I.

(iii') $\mu\delta_s^i = 2^{-l(s)-1}$.

(iv') For any q, the intervals δ_n^i of rank q are disjoint, and their unions is I.

(v') Each interval δ_n^i of rank less than q consists of a number of intervals of rank q.

Here we notice that if we decompose the sets E_n^i and δ_n^i of rank less than q as unions of the corresponding subsets and intervals of rank q (by (v) and (v')), then the indices occurring in these unions will be pairwise the same.

Fix $n = 2^{q+1}$. Because of properties (ii), (ii'), (iii) and (iii'), we can map each of the sets $E_k^i(\Lambda)$ of rank q onto the interval δ_k^i by means of a one-to-one measure-preserving piecewise linear function. Combining these mappings, we obtain, based on (iv) and (iv'), a one-to-one measure-preserving mapping $T_n^\Lambda : I \to I$. The properties (v) and (v') imply that

$$T_n^\Lambda(E_s^i) = \delta_s^i \quad (1 \leq s \leq n;\ i = 1, 2).$$

Finally, exploiting properties (i) and (i'), we convince ourselves of the relation

$$\chi_s^\Lambda(x) = \chi_s(T_n^\Lambda x) \quad (1 \leq s \leq n,\ x \in I).$$

Since q was arbitrary, we conclude that the systems χ^Λ and χ are weakly isomorphic.

This lemma shows that any finite collection of these polynomials, $\{\chi_k^\Lambda\}$, $1 \leq k \leq n$, is identical in its metric properties to a segment of the Haar system $\{\chi_k\}$, $1 \leq k \leq n$. In particular, the system χ^Λ is orthonormal. However, it is very incomplete. (The polynomials in χ^Λ are constructed exclusively from blocks of Haar functions that are spaced arbitrarily far apart, and each of these blocks defines only one of the functions χ_k^Λ.) Along this same line it can be shown that there exists a limit mapping

$$T^\Lambda = \lim_{n \to \gamma} T_n^\Lambda, \quad T^\Lambda : I \to I,$$

which, however, is quite noninvertible.

The property of the Haar system established here—the existence, in an arbitrarily sparse subsequence of blocks, of respective polynomials forming a system weakly isomorphic to the original system—determines the role that this system will play in investigation of the properties of arbitrary complete ONS.

With the applications of § 3 in mind, we shall prove the following theorem not only for complete ONS but also for bases in function spaces.

Let B be a separable Banach space of functions defined on $X = [0, 1]$, and suppose B satisfies the following set-theoretic inclusions:

$$L^\gamma \subset B; \qquad B^* \subset L. \tag{*}$$

The first of these conditions can be weakened: it is enough to require that B contain the set χ of all Haar polynomials (step functions with discontinuities at dyadic-rational points). The second inclusion is understood in the sense that the restriction to the set χ of any continuous linear functional $\Phi \in B^*$ can be written in the form

$$\Phi(f) \equiv (f, \Phi) = \int_X \alpha_\Phi(x)\, f(x)\, dx, \quad \alpha_\Phi \in L. \tag{6}$$

Theorem. *Let* $\phi = \{\phi_n\}$ *be an arbitrary basis in* B, *and let* $\psi = \{\psi_n\}$ *be the system of functionals dual to it. Then for any sequence* $\varepsilon_k > 0$ *it is possible to exhibit sequences of indices* $n_k \uparrow \infty$ *and of functions* $\hat\chi_k \in \chi$ *such that*

(i) *the systems* $\{\hat\chi_k\}$ *and* $\{\chi_k\}$ *are weakly isomorphic, and*

(ii) $\left\| \sum_{n=n_k+1}^{n_{k+1}} (\hat\chi_j, \psi_n) \phi_n \right\| < \varepsilon_j \varepsilon_k \quad (j, k = 1, 2, \ldots; j \neq k).$

We shall define by induction the sequence $\{n_k\}$ and an increasing sequence of natural numbers $\varLambda = \{\lambda_k\}$. The functions $\hat\chi_k$ will be defined by

$$\hat\chi_k = \chi_k^\varLambda. \tag{7}$$

Set $\lambda_1 = 0$, $\hat\chi_1 \equiv 1$, $n_1 = 0$, and, using the fact that the system ϕ is a basis, find a number $n_2 > n_1$ such that

$$\left\| \sum_{n=n_2+1}^{N} (\hat\chi_1, \psi_n) \phi_n \right\| < \varepsilon_1 \varepsilon_2 \quad (\forall N > n_2).$$

Suppose we have already defined the numbers $\lambda_1 < \cdots < \lambda_{s-1}$ and consequently (by (1), (2), (3) and (7)) also the polynomials $\hat\chi_k$ and the sets E_k^i, $i = 1, 2$; $1 \le k \le s-1$. Suppose also the numbers $n_1 < \cdots < n_s$ have been defined, and that for each $j = 1, 2, \ldots, s-1$ the following inequalities hold:

$$\left\| \sum_{n=n_k+1}^{n_{k+1}} (\hat\chi_j, \psi_n) \phi_n \right\| < \varepsilon_j \varepsilon_k \quad (1 \le k \le s-1, k \neq j);$$

$$\left\| \sum_{n=n_s+1}^{N} (\hat\chi_j, \psi_n) \phi_n \right\| < \varepsilon_j \varepsilon_s \quad (\forall N > n_s). \tag{8}$$

Consider the sequence of functions

$$\tilde\chi_{\lambda,s}(x) = 2^{(1/2)(l(s)-\lambda)} \chi(E_{\nu(s)}^{\alpha_s}; x) \sum_{r=1}^{2^\lambda} \chi_\lambda^{(r)}(x) \quad (\lambda > \lambda_{s-1}).$$

It is orthonormal on X. Indeed, we have from (4), (iii) and the definition of ν given with (3),

$$\int_X \tilde\chi_{\lambda,s}^2 dt = 2^{l(s)-\lambda} \cdot 2^\lambda \mu E_{\nu(s)}^{\alpha_s} = 1.$$

In checking orthogonality, it is convenient to write the polynomial $\tilde\chi_{\lambda,s}$ in the form given by (5). We have, for $p > q > \lambda_{s-1}$,

$$\int_X \tilde\chi_{p,s} \tilde\chi_{q,s} dx = 2^{l(s)-(1/2)(p+q)} \sum_{r=1}^{2^p} \sum_{m=1}^{2^q} \gamma_r(l) \gamma_m(q) \int_X \chi_p^{(r)} \chi_q^{(m)} dx = 0.$$

Further, the system $\{\tilde\chi_{\lambda,s}\}$ is uniformly bounded (for fixed s):

$$|\tilde\chi_{\lambda,s}(x)| \le 2^{(1/2)[l(s)-\lambda]} 2^{\lambda/2}.$$

Therefore by Lebesgue's theorem (see [1]) the Fourier coefficients of any integrable function with respect to this system tend to zero. Therefore, since $\tilde\chi_{\lambda,s} \in \chi$ and the functionals ψ_n restricted to χ can be written in the form of (6), we obtain that $(\tilde\chi_{\lambda,s}, \psi_n) = o_s(1)$. Based on this, we can choose a number $\lambda_s > \lambda_{s-1}$ so as to satisfy the inequalities

$$\left\| \sum_{n=n_k+1}^{n_{k+1}} (\tilde{\chi}_{\lambda_s,s}, \psi_n)\phi_n \right\| < \varepsilon_s \varepsilon_k \quad (1 \le k \le s-1). \tag{9}$$

By (1), (3), (2) and (7), when λ_s is defined, then $\hat{\chi}_s$ is also defined, and equals $\hat{\chi}_{\lambda_s,s}$. Expanding the vectors $\hat{\chi}_j$ with respect to the basis ϕ, we can find a number $n_{s+1} > n_s$ for which we have the estimate

$$\left\| \sum_{n=n_{s+1}+1}^{N} (\hat{\chi}_j, \psi_n)\phi_n \right\| < \varepsilon_j \varepsilon_{s+1} \quad (1 \le j \le s, N > n_{s+1}). \tag{10}$$

Inequalities (9) and (10) show that the induction assumptions (8) are justified at the s-th step. The induction thus constructs the sequences $\{n_s\}$ and Λ and the polynomials $\{\hat{\chi}_k\}$ in such a way that inequalities (8) are true for every $j < s$; $s = 1, 2, \ldots$. This justifies condition (ii) of the theorem. The lemma and the definition (7) ensure condition (i).

The following observations, which follow directly from this proof, will be used later.

1. We can take for $\{n_k\}$ in the theorem some subsequence of an arbitrary increasing sequence.

2. The mappings T_n that implement the weak isomorphism in the lemma are one-to-one on the set I and are made of piecewise linear functions.

§ 2. Divergent Fourier Series

In 1923 D. E. Menshov discovered that an orthogonal series from L^2 can be divergent. He exhibited [74] a special construction of an ONS $\{\phi_n\}$ such that for some sequence of coefficients $\{c_n\}$ satisfying the condition

$$\sum c_n^2 < \infty \tag{1}$$

the series

$$\sum c_n \phi_n(x) \tag{2}$$

diverges almost everywhere.

It was discovered further that a system with this property can be obtained from the classical trigonometric system by means of an appropriate rearrangement of its elements. The following theorem, stated by A. N. Kolmogorov in the article [61], is true: *there exists a function $f \in L^2[0,\pi]$ whose Fourier series $\sum c_n \cos nt$ after some rearrangement of terms diverges almost everywhere.*

Comparison of this result with Carleson's theorem, according to which a trigonometric Fourier series from L^2 in the natural ordering always converges almost everywhere, reveals the essential role of the manner of ordering of an ONS in questions of convergence of Fourier series. This fact acquires special importance for general ONS, which lack a natural ordering of their elements.

The following question arose in connection with Kolmogorov's theorem: does there exist a complete ONS of unconditional convergence (that is, one which is a system of convergence for any arrangement of its elements)? In other words, does there exist an ONS with respect to which every function $f \in L^2$ has a Fourier series converging to it almost everywhere for any ordering of the terms?

In 1960 Zahorskiĭ [168] published a brief outline of a proof of Kolmogorov's theorem. P. L. Ulyanov [152] showed that with appropriate modification this scheme can be applied to the Walsh system and—what is particularly important—to the Haar system, which are also not systems of unconditional convergence.

Using this, the author [90] and Ulyanov [153] gave a complete solution to the problem. The following theorem was proved.

Theorem 1. *For any complete ONS there exists a function $f \in L^2$ whose Fourier series* (2) *after some rearrangement of terms diverges almost everywhere.*

Thus the nature of the given phenomenon is not in the specific properties of the trigonometric system, but just in the completeness of the system. (There do exist incomplete systems of unconditional convergence, even bounded ones—for example, the Rademacher system.)

In the case of a uniformly bounded system ϕ, every series (2) with condition (1) can be made into a Fourier series from L^p $(p < \infty)$ by multiplying it termwise by a sequence of $\varepsilon_k = \pm 1$ (see Chap. II, § 2). Such a procedure does not affect unconditional convergence. Therefore for such systems the condition $f \in L^2$ in Theorem 1 can be replaced by $f \in \bigcap_{p < \infty} L^p$ [153].

Finally, in [91] there is a proof of the following.

Theorem 2. *For any complete ONS $\{\phi_n\}$, there exists a continuous function f whose Fourier series for some ordering of the terms diverges unboundedly almost everywhere.*

(A numerical series $\sum \alpha_k$ is said to diverge unboundedly if $\limsup\limits_{n \to \infty} \left| \sum\limits_1^n \alpha_k \right| = \infty$.)

The scheme of the proof of this theorem that we present below (see [91, 93]) is first to prove the result for the Haar system and then by the construction of § 1 to extend it to the general case.

ω-rearrangements of the Haar system. Our first goal is to construct a bounded function whose Fourier-Haar series diverges almost everywhere for some ordering of the terms. Such a construction for $f \in L^p$ $(p < \infty)$ was first carried out by P. L. Ulyanov [154] by means of a complicated construction based on Zahorskiĭ's method.

It was discovered further that for the Haar system one can exhibit a very simple rearrangement that allows one to construct divergent Fourier series. Consider the collection of Haar functions $\alpha_s = \{\chi_i\}$, $1 < i \leq 2^s$. To each i is associated the central zero t_i of the function χ_i, that is, the midpoint of the support of this function. It is easy to see that such a correspondence is one-to-one. Arrange the collection α_s to increase in order with the points t_i. We obtain as a result the rearranged collection $\omega \alpha_s = \{\chi_{\omega i}\}$ $(1 < i \leq 2^s)$. An interesting feature of this rearrangement is that it eliminates the interference in the Haar system. More precisely, for each $x \in [0,1]$ there clearly exists some number $\lambda = \lambda(x)$ for which the inequalities

$$\chi_{\omega_i}(x) \leq 0 \quad (1 < i \leq \lambda), \qquad \chi_{\omega_i}(x) \geq 0 \quad (\lambda < i \leq 2^s) \tag{3}$$

are true.

In this way the collection $\omega \alpha_s$ resembles a block in Menshov's system of divergence (see [55]).

The rearrangement ω was introduced in [91]. As Nikishin and Ulyanov showed later on [88], the following inequality is true for it:

$$\delta(x) = \max_{1 < l \leq 2^s} \left| \sum_{i=2}^{l} a_{\omega_i} \chi_{\omega_i}(x) \right| \geq \tfrac{1}{4} \sum_{1 < i \leq 2^s} a_i |\chi_i(x)| \quad (\forall \{a_i\}; x \in [0,1]). \tag{4}$$

Indeed, it follows from (3) that

$$\delta(x) \geq \tfrac{1}{2} \max \left\{ \left| \sum_{1 < i \leq \lambda(x)} a_{\omega_i} \chi_{\omega_i}(x) \right|, \left| \sum_{\lambda(x) < i \leq 2^s} a_{\omega_i} \chi_{\omega_i}(x) \right| \right\}$$

$$= \tfrac{1}{2} \max \left\{ \left| \sum_{1 < i \leq \lambda(x)} a_{\omega_i} |\chi_{\omega_i}(x)| \right|, \left| \sum_{\lambda(x) < i \leq 2^s} a_{\omega_i} |\chi_{\omega_i}(x)| \right| \right\} \geq \tfrac{1}{4} \sum a_i |\chi_i(x)|.$$

We shall prove the following lemma [91].

Lemma 1. *There exists a uniformly convergent series $\sum b_k \chi_k$ which after some rearrangement of its terms diverges unboundedly almost everywhere.*

Define

$$P_n(x) = \frac{1}{n^4} \sum_{k=n^8+1}^{n^8+n^7} \sum_{j=1}^{2^k} \varepsilon_k^{(j)} \frac{1}{\sqrt{2^k}} \chi_k^{(j)},$$

where the numbers $\varepsilon_k^{(j)} = 0, 1$ are determined by induction: at each step, if the sum of the terms already defined is < 1 for $x \in \Delta_k^{(j)} \equiv \left(\frac{j-1}{2^k}, \frac{j}{2^k} \right)$, then let $\varepsilon_k^{(j)} = 1$, and otherwise let $\varepsilon_k^{(j)} = 0$. Clearly $|P_n(x)| < 2$, and therefore, since the Lebesgue functions of the Haar system are uniformly bounded, we deduce that the series

$$\sum_n \frac{1}{n^6} \sum_{k=n^8+1}^{n^8+n^7} \sum_{j=1}^{2^k} \varepsilon_k^{(j)} \frac{1}{\sqrt{2^k}} \chi_k^{(j)} \tag{5}$$

converges uniformly. Letting $\delta_n(x)$ denote the maximum of the absolute values of the partial sums of the polynomial P_n, arranged in increasing order of the essential zeroes of the functions χ_i, we obtain from (4) that

$$\delta_n(x) \geq \frac{1}{4n^4} \sum_{k=n^8+1}^{n^8+n^7} \sum_{j=1}^{2^k} \varepsilon_k^{(j)} \frac{1}{\sqrt{2^k}} |\chi_k^{(j)}(x)|,$$

and further,

$$\delta_n(x) \geq \tfrac{1}{4} n^3 \quad \left(x \in I \setminus U_n, \; U_n = \bigcup_{j,k: \varepsilon_k^{(j)} = 0} \Delta_k^{(j)} \right). \tag{6}$$

Clearly $P_n(x) \geq 1$ $(x \in U_n)$. Therefore $\mu U_n \leq \|P_n\|_2 \leq \frac{1}{\sqrt{n}}$. Inequality (6) shows that if in series (5) we make the ω-rearrangement of the terms in the n-th block $(n = 1, 2, \ldots)$, then we obtain a series that diverges unboundedly on the set

$$E = \limsup \complement U_n,$$

which has full measure.

Proof of Theorem 2. Let $\{\phi_n\}$ be an arbitrary complete ONS in $L^2[0,1]$. Using the theorem of § 1 with $B=L^2$, we find an ONS $\{\hat{\chi}_j\}$ weakly isomorphic to the Haar system, and a sequence of indices $n_k\uparrow\infty$ giving the inequalities

$$\left\|\sum_{n=n_k+1}^{n_{k+1}}(\hat{\chi}_j,\phi_n)\phi_n\right\|<\frac{1}{2^{k+j}}\quad(j,k=1,2,\ldots;j\neq k).\tag{7}$$

Because of the isomorphism, the series $\sum b_j\hat{\chi}_j$ with the coefficients $\{b_j\}$ defined in Lemma 1 converges in the L^∞-metric to some function F and yet diverges unboundedly almost everywhere after a rearrangement of its terms. Consider the Fourier series

$$\sum c_n\phi_n,\qquad c_n=(F,\phi_n).\tag{8}$$

From (7) we have for $k=1,2,\ldots$ that

$$\sum_{n=n_k+1}^{n_{k+1}}c_n\phi_n=\sum_{n=n_k+1}^{n_{k+1}}\left(\sum_j b_j\hat{\chi}_j,\phi_n\right)\phi_n=\sum_{n=n_k+1}^{n_{k+1}}b_k(\hat{\chi}_k,\phi_n)\phi_n+\sum_{n=n_k+1}^{n_{k+1}}\sum_{j\neq k}b_j(\hat{\chi}_j,\phi_n)\phi_n;\tag{9}$$

$$\left\|\sum_{n=n_k+1}^{n_{k+1}}c_n\phi_n-b_k\hat{\chi}_k\right\|\leq|b_k|\left\|\sum_{n\notin[n_k+1,n_{k+1}]}(\hat{\chi}_k,\phi_n)\phi_n\right\|+\sum_{j\neq k}|b_j|\left\|\sum_{n=n_k+1}^{n_{k+1}}(\hat{\chi}_j,\phi_n)\phi_n\right\|$$

$$\leq\frac{\|F\|}{2^{k-1}}.$$

Hence it follows that

$$\sum_k\left|\sum_{n=n_k+1}^{n_{k+1}}c_n\phi_n(x)-b_k\hat{\chi}_k(x)\right|<\infty\quad\text{almost everywhere.}\tag{10}$$

Therefore series (8), written in the form $\sum_k\sum_{n=n_k+1}^{n_{k+1}}c_n\phi_n$, will, after some rearrangement of the terms in the outer summation, diverge unboundedly almost everywhere. Since $F\in L^\infty$, we see that the proof will be complete as soon as we establish the following general lemma.

Lemma 2. *Let $\{\phi_n\}$ be any ONS, and suppose the Fourier series of some function $F\in L^\infty$ diverges unboundedly almost everywhere on the set E. Then there exists a continuous function f having the same property.*

The function f is constructed by an inductive process according to the following scheme (for technical details see [93]). Assume that the numbers $n_1<n_2<\cdots<n_k$ and the function $f_k\in C$ are already defined in such a way as to satisfy the inequalities

$$\mu\left\{x\in E;\ \max_{1\leq l\leq n_s}\left|\sum_{i=1}^l c_i(f_k)\phi_i(x)\right|\leq s\right\}<\frac{1}{s}\quad(s=1,2,\ldots,k).\tag{11}$$

It is easy to see that for all λ, with the possible exception of a countable set, the function $f_k+\lambda F$ has a Fourier series that diverges unboundedly almost everywhere on E. Fix λ_{k+1} and a number n_{k+1} such that

$$\mu\left\{x\in E;\ \max_{1\leq l\leq n_{k+1}}\left|\sum_{i=1}^l c_i(f_k+\lambda_{k+1}F)\phi_i(x)\right|\leq k+1\right\}<\frac{1}{k+1}.\tag{12}$$

The number $\lambda_{k+1} < \frac{1}{2^{k+1}}$ can be chosen so small that inequality (11) remains valid for the function $F_{k+1} = f_k + \lambda_{k+1} F$.

We can now approximate F by a function G_{k+1}, $\|G_{k+1}\|_C \le \|F\|_\infty$, so that the first n_{k+1} Fourier coefficients change only a little, that is, preserving inequalities (11) and (12) for the function $f_{k+1} = f_k + \lambda_{k+1} G_{k+1}$. It is easy to see that $f = \lim_{k \to \infty} f_k$ satisfies the assertion of the lemma.

We remark that, as we shall show in Chap. IV, § 3, the conclusion of this lemma does not follow if we assume only that $F \in L^2$. This means, in particular, that the strengthening of Theorem 1 contained in Theorem 2 is not merely formal.

Generalizations. There have been many investigations of the possibility of generalizing Theorems 1 and 2 in various directions. We state some of the results that have been obtained.

1. The proof of Theorem 2 and Remark 1 of § 1 yield the following stronger proposition [93]: for any complete ONS $\{\phi_n\}$ and any increasing sequence of natural numbers $\{v_k\}$, it is possible to exhibit a function $f \in C$ whose Fourier series $\sum\limits_{k} \sum\limits_{v=v_k+1}^{v_{k+1}} c_v \phi_v$ diverges almost everywhere after some rearrangement of terms in the outer summation.

This means in particular that *if a complete ONS is divided into finite blocks of arbitrary length, then it is always possible to obtain a system of divergence by rearranging these blocks without changing the order of the functions within each block.* This variant of Theorem 1 will be used in Chap. IV.

2. There arises the question of whether it is possible in Theorem 1 to replace condition (1) (on the coefficients of a Fourier series that diverges after a rearrangement) by a stronger condition of the type

$$\sum c_n^2 \omega(n) < \infty, \tag{13}$$

where the sequence $\omega(n) \uparrow \infty$ is the same for all complete systems. A negative answer was given in [92]: *for any sequence $\omega(n) \uparrow \infty$ there exists a complete ONS for which condition (13) implies the absolute convergence of the series $\sum c_n \phi_{v_n}$ almost everywhere for all permutations $\{v_n\}$.*

Theorem 2 is likewise conclusive from the point of view of smoothness: the condition $f \in C$ cannot be replaced by any smoothness condition not depending on the system. Specifically, S. V. Bochkarev [15] showed that *for any modulus of continuity $\omega(\delta)$, there exists a complete ONS with respect to which every function $f \in H^\omega$ expands as a Fourier series absolutely convergent almost everywhere.*

However, these questions remain valid for individual systems (see § 5).

3. L. V. Taĭkov [133] using Theorem 2 showed that for the trigonometric system this theorem admits the following strengthening:

there exists a conjugate pair of continuous functions f, $\tilde{f}$ such that the trigonometric Fourier series of these functions both diverge at every point after some single rearrangement of the terms.

4. O. D. Tsereteli [150] showed that *for any complete system, the property that a function expands into an unconditionally convergent series is not a metric invariant.* More precisely: any complete ONS can be ordered in such a way that for any function $f \in L^2$, $f \not\equiv$ const, there is a metrically equivalent function $\hat{f}$ whose Fourier series diverges almost everywhere. This fact is due to Theorem 2 and the following general lemma (see [150]): if $\{\psi_n\}$ is an ONS with respect to which the Fourier series of some function $F \in L^\infty$, $\int_0^1 F dx \neq 0$, diverges unboundedly almost everywhere, then for any function $f \in L^2$, $f \not\equiv$ const, there exists a metrically equivalent function $\hat{f}$ whose Fourier series with respect to the system $\{\psi_n\}$ diverges unboundedly almost everywhere.

5. In connection with Theorem 1, this question arises: given a system of convergence $\{\phi_n\}$, which rearrangements are also systems of convergence, and which are not? This problem has been investigated only slightly. A. I. Rubinšteĭn [114] obtained some results for the classical systems. In particular, he proved that *for any sequence $\mu_k \uparrow \infty$, $\mu_k = o(k)$, there exists a permutation $\{n_k\}$ of the natural numbers such that $\dfrac{n_k}{k} - 1 = o\left(\dfrac{1}{\mu_k}\right)$ and such that the system $\{\cos n_k x\}$ is a system of divergence.*

6. The basic results extend to summability methods. Specifically, in Theorem 2 it may be asserted that the Fourier series after an appropriate rearrangement is not summable by any pre-assigned linear Toeplitz method. The possibility of such a generalization is due to general results on the connection between unconditional summability and convergence of Fourier series [55]. For a discussion of this connection for arbitrary series of functions, see [95].

7. We note, finally, some generalizations concerning nonorthogonal systems. Suppose a Banach space B of functions satisfies condition (*) (§ 1) and the following:

$$\|f\|_{L_\mu} \leq \|f\|_B \leq \|f\|_{L^\infty} \quad (\forall f \in B) \tag{14}$$

(μ being some measure).

Then the following proposition is true: for any basis $\{\phi_n\}$ in B, there exists a function $f \in C$ whose Fourier series

$$\sum c_n \phi_n, \quad c_n = (f, \psi_n) \quad ((\phi_n, \psi_n) = \delta_{nk})$$

after some rearrangement diverges unboundedly almost everywhere with respect to the measure μ.

The proof is the same as for Theorem 2. Here the right-hand inequality in (14) ensures the convergence of the series $\sum b_k \hat{\chi}_k$ in the metric of B, and the left-hand inequality allows us to deduce (10) from relation (9).

After this one must use the appropriate variant of Lemma 2.

The special case $B = L^2$, $f \in L^2$ was studied by P. L. Ulyanov [154]; the case $B = L^p$, $f \in L^p$, $1 < p < \infty$, was studied by F. G. Arutyunyan [3], who used the method of [93]. See also [100].

We mention also the work of A. A. Talalyan [136] and F. A. Talalyan [140], where similar problems are considered for representation systems in certain spaces. This means that the condition that $\{\phi_k\}$ be a basis is replaced by the weaker requirement that every vector f be representable, possibly non-uniquely, as a series $\sum c_k \phi_k$.

§ 3. Bases in Function Spaces and Majorants of Fourier Series

The Haar system is the best basis. Let B be a separable Banach space and let $\{\phi_i\}$ be a closed system of elements in B. As is known, a necessary and sufficient condition for this system to be a basis is the finiteness of the Banach constant

$$K(\phi) = \sup_{p \in P_\phi, \|p\| \le 1} \max_{1 \le l \le n} \left\| \sum_{i=1}^{l} a_i \phi_i \right\|,$$

where $P_\phi = \{p\}$ is the set of all linear combinations $p = \sum_{i=1}^{n} a_i \phi_i$, $n = 1, 2, \dots$.

Suppose the Haar system is closed in a Banach space B that consists of measurable functions on $[0,1]$ and satisfies the condition that if $f \in B$, then every function $\hat{f}$ metrically equivalent to f also belongs to B, and

$$\|\hat{f}\| = \|f\|. \tag{**}$$

Some special cases are the L^p-spaces for $1 \le p < \infty$, the separable Orlicz spaces, etc.

It is not difficult to convince oneself that under these conditions *the Haar system has the smallest possible Banach constant:* $K(\chi) = 1$.

Indeed, let $p = \sum_{i=1}^{2^k} a_i \chi_i$, $2^s < l \le 2^{s+1}$, $s < k$. On each of the intervals $\varDelta$ of constancy of the functions $\{\chi_i\}$ $(1 \le i \le l)$ define an "addition": $x \dotplus h = y$, for x and $y \in \varDelta$, when $x + h - y \equiv 0 \bmod |\varDelta|$. Define

$$p^{(v)}(x) = p\left(x \dotplus \frac{v}{2^k}\right).$$

It is easy to see that

$$\sum_{i=1}^{l} a_i \chi_i(x) = \frac{1}{|\varDelta|} \int_{\varDelta} p(t) dt = \frac{1}{2^{k-s}} \sum_{v=1}^{2^{k-s}} p^{(v)}(x), \quad x \in \varDelta \cap I.$$

Moreover, the functions $p^{(v)}$ are metrically equivalent to p. Therefore the triangle inequality and condition (**) give

$$\left\| \sum_{i=1}^{l} a_i \chi_i \right\| \le \|p\|,$$

as required.

From this result in particular it follows that *the Haar system forms a basis in the space B*.

In a number of problems of Banach space geometry, the concept of an *unconditional basis* plays an important role (see [26, 68]). A system ϕ is called an unconditional basis if it is a basis under every ordering. For example, any complete ONS is an unconditional basis in L^2. However, the following result, due to Orlicz, is true: *a uniformly bounded ONS cannot be an unconditional basis in an L^p-space for any $p \neq 2$.* (In particular, the trigonometric system is a conditional basis in the L^p-spaces for $1 < p < \infty$, $p \neq 2$.)

Indeed, it is sufficient to examine the case $p < 2$ (see Chap. IV, § 4). As Orlicz [109] showed, unconditional convergence of the series $\sum f_k$ in L^p, $p < 2$, implies that $\sum \|f_k\|_p^2 < \infty$. For a series $f = \sum c_k \phi_k$ with respect to a uniformly bounded ONS, this is equivalent to the condition $\sum c_k^2 < \infty$, which can be true only for $f \in L^2$.

For a generalization to nonorthogonal bases see V. F. Gaposhkin [36].

The Haar system, however, *does form an unconditional basis in L^p for all $p \in (1, \infty)$* (Marcinkiewicz [72]). A key role in the proof is played by Paley's inequality

$$C_p' \left\| \sqrt{\sum_1^n a_k^2 \chi_k^2} \right\|_p \leq \left\| \sum_1^n a_k \chi_k \right\|_p \leq C_p \left\| \sqrt{\sum_1^n a_k^2 \chi_k^2} \right\|_p, \tag{1}$$

where C_p' and C_p are positive constants depending only on p. It follows from this inequality that for any Haar polynomial and for any collection of numbers $\gamma_k = \pm 1$ the relation

$$\left\| \sum_1^n \gamma_k a_k \chi_k \right\|_p \leq \frac{C_p}{C_p'} \left\| \sum_1^n a_k \chi_k \right\|_p$$

holds, and one has then to use the following general criterion: a basis $\{\phi_k\}$ in a Banach space is unconditional if and only if the quantity

$$\tilde{K}(\phi) = \sup_{\{\gamma_k = \pm 1\}} \sup_{p \in P_\phi \cdot \|p\| = 1} \left\| \sum_1^n \gamma_k a_k \phi_k \right\|$$

is finite.

It turns out that in a wide class of function spaces the Haar system has the best unconditional constant $\tilde{K}(\phi)$.

Theorem 1 [100]. *Let B be a Banach space satisfying conditions (*) (§ 1) and (**). Then any basis ϕ satisfies the inequality*

$$\tilde{K}(\phi) \geq \tilde{K}(\chi). \tag{2}$$

Finiteness of these quantities is not assumed here. The proof is based on the results of § 1. Fix arbitrary numbers $\mu < K(\chi)$ and $\delta > 0$, and find a polynomial $p = \sum_1^l a_k \chi_k$, with $\|p\| = 1$, and a collection of numbers $\{\gamma_k = \pm 1\}$ satisfying the inequality

$$\left\| \sum_{k=1}^l \gamma_k a_k \chi_k \right\| > \mu. \tag{3}$$

Next, on the basis of the theorem of § 1, we find a sequence $\{\tilde{\chi}_k\}$ weakly isomorphic to the Haar system, and numbers $\{n_k \uparrow\}$ for which the inequality

$$\left\| \sum_{n=n_j+1}^{n_{j+1}} (\tilde{\chi}_k, \psi_n) \phi_n \right\| < \frac{\varepsilon}{2^j} \quad (\forall k, j; k \neq j) \tag{4}$$

is true, where ψ is the system dual to the basis ϕ, and $\varepsilon < \delta \left(\sum |a_k| \right)^{-1}$.

It follows from the weak isomorphism that the polynomials $\sum_{k=1}^{l} \alpha_k \chi_k$ and $\sum_{k=1}^{l} \alpha_k \hat{\chi}_k$ are metrically equivalent, and this together with (**) implies

$$\left\| \sum_{1}^{l} \alpha_k \chi_k \right\| = \left\| \sum_{1}^{l} \alpha_k \hat{\chi}_k \right\| \quad (\forall \{\alpha_k\}). \tag{5}$$

Furthermore, we have, from (4),

$$\hat{\chi}_k = \sum_{n} (\hat{\chi}_k, \psi_n) \phi_n = \sum_{n=n_k+1}^{n_{k+1}} (\hat{\chi}_k, \psi_n) \phi_n + \rho_k; \quad \|\rho_k\| < \varepsilon. \tag{6}$$

Consider this polynomial in the system ϕ:

$$\tilde{P} = \sum_{k=1}^{l} a_k \sum_{n=n_k+1}^{n_{k+1}} (\hat{\chi}_k, \psi_n) \phi_n.$$

Inequalities (5) and (6) give

$$\|\tilde{P}\| = \left\| \sum_{1}^{l} a_k \hat{\chi}_k \right\| + \sum_{1}^{l} |a_k| \|\rho_k\| \leq 1 + \delta.$$

Using these inequalities again in conjunction with (3), we obtain that

$$\left\| \sum_{k=1}^{l} \gamma_k a_k \sum_{n=n_k+1}^{n_{k+1}} (\hat{\chi}_k, \psi_n) \phi_n \right\| \geq \left\| \sum_{1}^{l} \gamma_k a_k \hat{\chi}_k \right\| - \delta > \mu - \delta,$$

whence, by the choice of μ and δ, (2) follows.

The theorem just proved immediately implies

Corollary [100]. *Suppose the Haar system is closed in a space B satisfying conditions* (*) *and* (**). *Then either* $\{\chi_k\}$ *forms an unconditional basis in B, or there does not exist any unconditional basis at all in this space.*

This result reduces the problem of the existence of an unconditional basis in the space B to the problem of whether the Haar system is such a basis.

Let us note some concrete applications.

1. It is easy to see that the Haar system is not an unconditional basis in the space $L[0,1]$. This fact is a consequence of the following easily verified relations:

$$\left\| \chi_0^{(0)} + \chi_0^{(1)} + \sum_{s=1}^{2k} \sqrt{2^s} \chi_s^{(1)} \right\|_1 = 1; \qquad \left\| \chi_0^{(1)} + \sum_{\sigma=1}^{k} \sqrt{2^{2\sigma}} \chi_{2\sigma}^{(1)} \right\|_1 > \frac{k}{4}.$$

Hence, by the preceding corollary, we obtain a theorem first proved by Pelczynski [112] by another method: *there does not exist an unconditional basis in the space* $L[a,b]$.

An analogous result holds for the space $C[a,b]$ (Karlin). This case does not fit formally into the scheme outlined above, as the conditions (*) and (**) are violated, but the same approach with some modifications (see [5]) makes it possible to prove this theorem, too.

2. Pelczynski, using Theorem 1, exhibited a separable reflexive space with no unconditional basis. This property is enjoyed by the space $B = \bigoplus_{l_2} L^{p_n}, 1 < p_n \downarrow 1$, whose elements are all the sequences $x = \{x_n\}, x_n \in L^{p_n}$, that have finite norm $\|x\| = (\sum \|x_n\|_{p_n}^2)^{1/2}$.

This result has subsequently been surpassed: P. Enflo [31], in response to the famous problem of Banach, constructed an example of a separable (and reflexive) space without a basis.

3. We note some applications to the theory of Orlicz spaces. (For the definition and properties of these spaces see [65].) The separable Orlicz spaces are known to satisfy conditions (*) and (**).

Chen [167] showed that Paley's inequality (1) is valid in reflexive Orlicz spaces. It follows from this (see [16, 38]) that the Haar system forms an unconditional basis in these spaces.

V. F. Gaposhkin established [38] that the condition of reflexivity is also a necessary condition for the Haar system to be an unconditional basis in an Orlicz space. It then follows from the Corollary to Theorem 1 that *in the nonreflexive Orlicz spaces there do not exist any unconditional bases.*

4. E. M. Semenov studied [118] the general case of symmetric spaces. These are spaces B of measurable functions satisfying the conditions

(i) if $y \in B$ and $|x(t)| \le |y(t)|$, then $x \in B$ and $\|x\| \le \|y\|$;

(ii) if $|x|$ and $|y|$ are equimeasurable and $x \in B$, then $y \in B$ and $\|x\| = \|y\|$.

The fundamental function $\phi(\tau) = \|\chi([0, \tau]; t)\|$ is important in characterizing these spaces.

Using an interpolation theorem that he found in symmetric spaces, Semenov showed that the Haar system is an unconditional basis in a separable symmetric space B if and only if

$$1 < \liminf_{t \to 0} \frac{\phi(2t)}{\phi(t)} \le \limsup_{t \to 0} \frac{\phi(2t)}{\phi(t)} < 2.$$

On the basis of the Corollary of Theorem 1 he then concluded that *this condition is necessary and sufficient for the existence of an unconditional basis in B.*

In connection with what was said at the beginning of this section, it is of interest to investigate whether the Haar system is a basis in function spaces with a norm not invariant with respect to measure-preserving transformations of the interval $[0, 1]$. Examples of such spaces are the weighted spaces $L_g^p, \|f\| = \left(\int_0^1 |f|^p g \, dt\right)^{1/p}$, where $0 < g(t) \in L$ is a fixed weight. A. S. Krantsberg showed [64] that the following condition is necessary and sufficient for $\{\chi_n\}$ to be a basis in L_g^p $(1 \le p < \infty)$:

$$\sup_n \left[\|\chi_n\|_{L_g^p} \left\|\frac{\chi_n}{g}\right\|_{L_g^q}\right] < \infty \quad \left(q = \frac{p}{p-1}\right).$$

It would be interesting to find conditions on the weight which would make the Haar system an unconditional basis in L_g^p spaces.

Majorants of the partial sums of Fourier series. In investigation of the convergence of Fourier series

$$f \sim \sum c_k \phi_k, \tag{7}$$

the following quantity plays an important role:

$$\delta_\phi(f; x) = \sup_n \left| \sum_{k=1}^n c_k \phi_k(x) \right|.$$

Most theorems about the convergence almost everywhere of series (7) actually make some estimates of the majorant $\delta(f)$. For example, the Menshov-Rademacher theorem (Chap. II) is supplemented by the following inequality (L. V. Kantorovich; see [1] (Chap. II, § 3)):

$$\|\delta(f)\|_2 \leq K \left[\sum_1^\infty c_n^2 \ln^2(n+1) \right]^{1/2}.$$

Further, if $\phi = \{\phi_n\}$ is a system of convergence, then for any function $f \in L^2$ we have $\delta(f; x) < \infty$ almost everywhere; it is easy to see that the converse is also true. As a rule, in this case there is the stronger inequality $\|\delta\|_2 \leq K \|f\|_2$.

It is clear that if a system ϕ satisfies the inequality

$$\|\delta(f)\|_p \leq K_p \|f\|_p \quad (\forall f \in L^p), \tag{8}$$

then this system (if it is closed) forms a basis in L^p.

Here are some examples of this. The trigonometric system forms a basis in L^p, $1 < p < \infty$ (M. Riesz), and is a system of convergence almost everywhere in these classes (Carleson, Hunt). Both of these facts are contained in the following assertion, proved by Hunt [53]: *the trigonometric system satisfies inequality* (8) *for every* $p \in (1, \infty)$. For the Haar system, this result was established by Marcinkiewicz. It is a consequence of the Hardy-Littlewood maximal theorem, according to which the quantity

$$\theta(f; x) = \sup_{\substack{0 \leq \xi \leq 1 \\ \xi \neq x}} \left| \frac{1}{\xi - x} \int_\xi^x f(t) \, dt \right|$$

satisfies the inequality

$$\|\theta(f)\|_p \leq K_p \|f\|_p \quad (\forall f \in L^p, 1 < p < \infty).$$

It is enough to observe that

$$\delta_\chi(f; x) = \sup_{\substack{k, j, \Delta_k^{(j)} \\ x \in \Delta_k^{(j)}}} \frac{1}{|\Delta_k^{(j)}|} \left| \int_{\Delta_k^{(j)}} f(t) \, dt \right| \leq 2 \theta(f; x). \tag{9}$$

For $p = 1$, inequality (8) is not true of the trigonometric system, as can be deduced simply from the fact that this system is not a basis in L. For the Haar system inequality (8) is also false for $p = 1$ [155]. Indeed, if $f \geq 0$ is monotonic, then we clearly have the estimate that is the reverse of (9), $\theta(f; x) \leq 2 \delta_\chi(f; x)$; at the same time, there exists $f_0 \in L$, $0 \leq f_0(x) \downarrow$, for which $\theta(f_0) \notin L$; for example, $f_0(t) = (t \ln^2 t)^{-1}$, $0 < t < 1/2$ (see [171] Chap. I).

The results of § 1 permit us to conclude from this that for $p=1$ the situation is similar for any complete system. In view of what has been said above, it makes sense to study just bases in L.

Theorem 2. *Let $\{\phi_n\}$ be an arbitrary basis in the space $L[0,1]$. Then there exists $f \in L$ for which $\delta(f) \notin L$.*

Using the theorem of § 1, we find a sequence $\{\hat\chi_k\}$ weakly isomorphic to the Haar system and numbers $n_k \uparrow \infty$ such that the inequalities

$$\left\| \sum_{n=n_k+1}^{n_{k+1}} (\hat\chi_j, \psi_n)\phi_n \right\|_1 < \frac{1}{2^{k+j}} \quad (k \neq j)$$

are satisfied, where ψ is the system dual to the basis ϕ.

The series $\sum b_k \hat\chi_k$, where $\{b_k\}$ are the Fourier-Haar coefficients of the function f_0 (see above), converges in the L-metric to some function f.

Let $T_v: I \to I$ be an invertible measure-preserving transformation such that $\hat\chi_k(x) = \chi_k(T_v x)\,(1 \leq k \leq v)$. Then we have

$$\|\delta_{\hat\chi}(f)\| = \int_0^1 \sup_n \left| \sum_{k=1}^n b_k \hat\chi_k(x) \right| dx = \lim_{v \to \infty} \int_0^1 \sup_{1 \leq n \leq v} \left| \sum_{k=1}^n b_k \chi_k(T_v x) \right| dx$$

$$= \lim_{v \to \infty} \int_0^1 \sup_{n \leq v} \left| \sum_1^n b_k \chi_k(x) \right| dx = \|\delta_\chi(f_0)\| = \infty. \tag{10}$$

Let $c_n = (f, \psi_n)$, and make an estimate analogous to (9) of § 2. Since $b_k = O(2^{k/2})$, we hereby obtain

$$\sum_k \left\| \sum_{n=n_k+1}^{n_{k+1}} c_n\phi_n - b_k\hat\chi_k \right\| < \infty,$$

whence with the help of (10) we conclude

$$\|\delta_\phi(f)\| \geq \left\| \sup_l \sum_{k=1}^l \sum_{n=n_k+1}^{n_{k+1}} c_n\phi_n(x) \right\|$$

$$\geq \|\delta_{\hat\chi}(f)\| - \left\| \sum_k \left| \sum_{n=n_k+1}^{n_{k+1}} c_n\phi_n(x) - b_k\hat\chi_k(x) \right| \right\| = \infty,$$

as was required.

Observe that $f_0 \in L(\ln^+ L)^{1-\varepsilon}$; that is, $\int_X |f|(\ln^+ |f|)^{1-\varepsilon} dx < \infty$. Actually the following assertion is true (see Stein [129]): $f \in L\ln^+ L \Leftrightarrow \delta_f \in L$. This permits a strengthening of Theorem 2, by requiring some condition just slightly weaker than $f \in L\ln^+ L$; for example, $f \in L(\ln^+ L)^{1-\varepsilon}$, $\varepsilon > 0$. The result fails for $\varepsilon = 0$, as the example of the Haar system shows.

This same method shows that the constant $K_p(\phi)$ in inequality (8) for each $p > 1$ attains its minimum value in the case $\phi = \chi$. An analogous result is true also for other Banach spaces satisfying conditions (*) and (**).

We note also the following inequality for the Haar system:

$$K_1 \|[\sum c_n^2(f)\chi_n^2]^{1/2}\|_1 \leq \|\delta_\chi(f)\|_1 \leq K_2 \|[\sum c_n^2(f)\chi_n^2]^{1/2}\|_1 \quad (K_i > 0). \tag{11}$$

The proof (see [17, 25]) involves probability-theoretic ideas and uses the fact that the partial sums of any series $\sum c_n \chi_n$ are a martingale.

From (11) it follows in particular that the condition $\delta_\chi(f) \in L$ is necessary and sufficient for the unconditional convergence of a Fourier-Haar series in the L-metric. Indeed, the unconditional convergence implies, according to a theorem of Orlicz [47], the finiteness of the right-hand side of the inequality. In turn, if $\delta_\chi(f) \in L$, (11) shows that the majorant of any subseries is integrable, whence unconditional convergence in L follows.

Returning to the connection between convergence almost everywhere and properties of the majorants of Fourier series, we mention that some general results of E. Stein [128] under a certain assumption about the system ϕ justify the following proposition:

if the Fourier series of every function $f \in L^p$ $(1 \le p \le 2)$ converges almost everywhere, then the operator $\delta: f \mapsto \delta_\phi(f)$ is of weak type (p, p) (see [171]); in particular, we have the inequality

$$\|\delta(f)\|_{p-\varepsilon} \le K_{p,\varepsilon} \|f\|_p \qquad (\varepsilon > 0).$$

The assumption referred to is that the operators $S_n(f) = \sum_1^n c_k(f) \phi_k$ commute with translations (or with some ergodic family of measure-preserving transformations of the interval—Sawyer [117]). In particular, this assumption is satisfied by the trigonometric system under any ordering, and by the Walsh system. However, it is hopelessly violated for a rearrangement of the Haar system. There is, based on this, an example (see [98]) of a complete uniformly bounded ONS having the following properties:

(i) for any function $f \in L^2$, the Fourier series converges to f almost everywhere, and

(ii) for some $f \in C$, the majorant $\delta(f; x) \notin \bigcup_{p>0} L^p$.

It turns out, however, that the assertion of Stein's theorem does hold in the general case if sets of small measure are disregarded. Namely, it follows from results of E. M. Nikishin [84, 86] that for any ONS on $[a, b]$ that is a system of convergence almost everywhere in the class L^p $(1 \le p \le 2)$, it is possible to remove from the interval $[a, b]$ a set E of arbitrarily small measure and thereby make the operator δ to be of weak type $L^p[a, b] \to L^p(\complement E)$.

§ 4. Fourier Coefficients of Continuous Functions

As is well known, if a function $f \in L^q$, $1 \le q \le 2$, then its Fourier coefficients with respect to the trigonometric system (or any uniformly bounded ONS) satisfy the condition $\sum |c_n|^p < \infty$ $\left(p = \dfrac{q}{q-1} \right)$ (Hausdorff-Young-Riesz).

For $q > 2$, the corresponding statement is not true. Improving the properties of the function all the way to continuity still does not make the Fourier coefficients decrease more rapidly. This fact was established by Carleman, who showed that there exists a continuous function f whose Fourier coefficients with respect to the trigonometric system satisfy the condition

$$\sum |c_n|^p = \infty \qquad (\forall p < 2). \tag{1}$$

In this connection one says that a continuous function f has the *Carleman singularity* with respect to the ONS $\{\phi_n\}$ ($f \in Cr(\phi)$) if the Fourier coefficients $c_n^\phi(f) = (f, \phi_n)$ satisfy condition (1).

Carleman's theorem was extended to bounded systems by Orlicz (see [55] Chap. VI) and to arbitrary complete ONS by the author [94]. The latter result, roughly speaking, means that in the space L^2 it is impossible to choose a coordinate system in which the continuous functions would be expressed any better than the others. This result, in particular, is a consequence of Theorem 2 of § 2 and the fact that the condition $\sum |c_n|^p < \infty$ ($p < 2$) implies the unconditional convergence almost everywhere of an orthogonal series (Menshov).

We shall say that a function $f \in C$ has a *local Carleman singularity* with respect to the system ϕ ($f \in Cr^{loc}(\phi)$) if there is some point $t_0 \in [a, b]$ such that every function $F \in L^2$ coinciding with f in a neighborhood of this point satisfies the condition $\sum |c_n^\phi(F)|^p = \infty$ ($\forall p < 2$).

For the trigonometric system, the Carleman singularity is always local (Wiener). This situation is typical, and is true also for the other classical systems. In the general case there is

Theorem 1 (see [99]). *For any complete ONS there exists a continuous (even differentiable) function having a local Carleman singularity.*

However there do exist systems for which not only local but also nonlocal singularities are possible (see about this below).

It turns out that *a local Carleman singularity is always determined by the values of the function on some compact set of measure zero.*

The following proposition is true.

Theorem 2. *Let $\{\phi_n\}$ be an arbitrary complete ONS in $L^2[a, b]$, and suppose $f \in Cr^{loc}(\phi)$. Then there exists a compact set $\mathcal{K} \subset [a, b]$, $\mu \mathcal{K} = 0$, such that every function $F \in C[a, b]$ satisfying $F(x) = f(x)$ for all $x \in \mathcal{K}$ has a (local) Carleman singularity.*

Hence because of Theorem 1 it follows that

Theorem 3. *For any complete ONS it is possible to exhibit a function continuous on some compact set $\mathcal{K}$ of measure zero such that every continuous continuation of this function from $\mathcal{K}$ to the whole interval $[a, b]$ has the Carleman singularity.*

Thus the values of a continuous function on a closed set of measure zero, which is negligible in the calculation of the Fourier coefficients, nevertheless make a decisive contribution to their behavior as a whole.

We introduce the notation

$$\|f\|_{p, \phi} = \left(\sum_k |c_k^\phi(f)|^p \right)^{1/p}; \quad \|f\|_{p, \phi, \Delta} = \inf \|\tilde{f}\|_{p, \phi}. \tag{2}$$

Here $\varDelta$ is an interval contained in $[a,b]$, and the infimum is taken over all $\tilde{f} \in L^2$ such that $\tilde{f} \equiv f$ on $\varDelta$.

Theorem 2 is deduced from the following proposition.

Basic Lemma. *Suppose a complete ONS* $\{\phi_n\}$ *and* $\varDelta \subset [a,b]$ *are given. Then for any function* $f \in C$ *and any numbers* $p \in [1,2)$ *and* $v < \|f\|_{p,\phi,\varDelta}$, *there exists a compact set* $\mathscr{K} \subset \varDelta$, $\mu \mathscr{K} = 0$, *such that whenever a function* $F \in L^2$ *satisfies the conditions*

$$F|_{\varDelta} \in C, \qquad F|_{\mathscr{K}} = f|_{\mathscr{K}}, \tag{3}$$

then

$$\|F\|_{p,\phi} \geq v. \tag{4}$$

Here we allow the possibility that $\|f\| = \infty$. The proof is based on the construction of § 1.

For a given compact set $\mathscr{K}$, let us denote by $\mathscr{K}^s$ the union of all the intervals $\varDelta_s^{(j)} = \left[\dfrac{j-1}{2^s}, \dfrac{j}{2^s} \right]$ that have a nonempty intersection with $\mathscr{K}$.

Lemma 1. *Suppose numbers* $v, \varepsilon > 0$ *and* $p \in [1,2)$ *are given. Then there exists a compact set* $\mathscr{K} \subset [0,1]$, $\mu \mathscr{K} = 0$, *such that whenever* $f \in L$ *satisfies the conditions*
   (i) $|f(x)| < \varepsilon$ ($\forall x \in \mathscr{K}^s$) *for some* s, *and*
   (ii) $\|f\|_{p,\chi} < v$ (χ *denoting the Haar system*),
then

$$\left| \int_0^1 f \, dx \right| < 2\varepsilon. \tag{5}$$

Fix numbers $n_k \uparrow$ such that

$$n_0 = 0, \qquad \left(\frac{1}{p} - \frac{1}{2} \right)(m_k - 1) > \tfrac{3}{2} n_k - \log_2 \frac{\varepsilon}{2v}, \qquad m_k = n_{k+1} - n_k. \tag{6}$$

Define

$$E_0 = \varDelta_0^{(1)}, \qquad E_s = \bigcup_{j=1}^{2^{n_s-1}} \varDelta_{n_s}^{(2j-1)} \quad (s > 0); \qquad \mathscr{K} = \bigcap_{s=0}^{\infty} E_s.$$

$\mathscr{K}$ is clearly compact, and

$$\mathscr{K}^{n_s} = \bigcap_{i=0}^{s} E_i; \qquad \mu \mathscr{K}^{n_s} = \frac{1}{2^s};$$

so $\mu \mathscr{K} = 0$. Denote by σ_s the set of intervals $\varDelta_{n_s}^{(j)}$ composing the set $\mathscr{K}^{n_s}$. Suppose the function f satisfies condition (ii). Define

$$M(f;\varDelta) = \frac{1}{|\varDelta|} \int_{\varDelta} f \, dx; \qquad M_s(f) = \max_{\varDelta \in \sigma_s} |M(f;\varDelta)|.$$

We shall verify the following inequality:

$$M_s(f) \leq M_{s+1}(f) + \frac{\varepsilon}{2^{s+1}} \quad (s = 0, 1, \ldots). \tag{7}$$

Suppose we have fixed s and the interval $\Delta \equiv \Delta_{n_s}^{(j)} \in \sigma_s$. Define

$$\delta_i = \Delta_{n_s+1}^{(j-1)2^{m_s}+2i-1}, \quad \tilde{\delta}_i = \Delta_{n_s+1}^{2^{m_s}(j-1)+2i} \quad (1 \leq i \leq 2^{m_s-1}).$$

Manifestly,

$$\delta_i \in \sigma_{s+1}; \quad \bigcup_{i=1}^{2^{m_s-1}} (\delta_i \cup \tilde{\delta}_i) = \Delta. \tag{8}$$

We have further that

$$c_k^{(i)} = \int_0^1 f \cdot \chi_{n_s+1-1}^{(j-1)2^{m_s-1}+i} dt = (2|\delta_i|)^{-1/2} \left(\int_{\delta_i} f\, dt - \int_{\tilde{\delta}_i} f\, dt \right)$$

$$= (|\delta_i|/2)^{1/2} [M(f;\delta_i) - M(f;\tilde{\delta}_i)],$$

whence

$$v > \left(\sum_{n=1}^{\chi} |c_n^\chi(f)|^p \right)^{1/p} \geq \left(\sum_{i=1}^{2^{m_s-1}} |c_k^{(i)}|^p \right)^{1/p}$$

$$= 2^{-(n_s+1+1)/2} \left(\sum_{i=1}^{2^{m_s-1}} |M(f;\delta_i) - M(f;\tilde{\delta}_i)|^p \right)^{1/p}. \tag{9}$$

Next, because of (8) we have

$$M(f;\Delta) = \frac{1}{|\Delta|} \left[\sum_i \int_{\delta_i \cup \tilde{\delta}_i} f\, dt \right] = 2^{-m_s} \sum_i [M(f;\delta_i) + M(f;\tilde{\delta}_i)] \tag{10}$$

$$= 2^{1-m_s} \sum_{i=1}^{2^{m_s-1}} M(f;\delta_i) + 2^{-m_s} \sum [M(f;\tilde{\delta}_i) - M(f;\delta_i)] \equiv \Sigma_1 + \Sigma_2.$$

From (9) and (6) we obtain

$$|\Sigma_2| \leq 2^{-m_s} \left[\sum_{i=1}^{2^{m_s-1}} |M(f;\tilde{\delta}_i) - M(f;\delta_i)|^p \right]^{1/p} \cdot (2^{m_s-1})^{1-1/p} \tag{11}$$

$$< v\, 2^{n_s/2 + (m_s-1)(1/2-1/p)} < \varepsilon/2^{s+1}.$$

It is evident from (8) that $|\Sigma_1| \leq M_{s+1}(f)$, and this along with (10) and (11) gives

$$|M(f;\Delta)| \leq M_{s+1}(f) + \varepsilon/2^{s+1},$$

which proves inequality (7). Suppose the function f satisfies condition (i) for $s = s_0$. Since $n_s \geq s$, we shall have that $M_{s_0}(f) \leq \varepsilon$, whence, by successive applications of (7), we derive

$$\left| \int_0^1 f\, dx \right| = M_0(f) < M_1(f) + \frac{\varepsilon}{2} < M_2(f) + \frac{\varepsilon}{2} + \frac{\varepsilon}{4} < \cdots < M_{s_0}(f) + \varepsilon \sum_1^{s_0} 2^{-s} < 2\varepsilon.$$

Lemma 2. *Suppose we are given an increasing sequence of natural numbers* $\Lambda = \{\lambda_k\}$, *numbers* v, $\varepsilon > 0$, *and* $p \in [1,2)$. *Then there exists a compact set* $\mathcal{K}_\Lambda(\varepsilon, v, p) \subset [0,1]$, $\mu \mathcal{K}_\Lambda = 0$, *such that whenever* $f \in L[0,1]$ *satisfies the conditions* $|f(x)| < \varepsilon$ *on some neighborhood of* $\mathcal{K}_\Lambda$ *and* $\|f\|_{p,\chi^\Lambda} < v$, *where* χ^Λ *is the system defined in § 1, then condition (5) holds.*

According to § 1 (the lemma and Remark 2), for each l there exists a one-to-one piecewise linear measure-preserving mapping $T_l: I \rightarrow I$ such that

$$\chi_k^A(x) = \chi_k(T_l x) \quad (x \in I, 1 \le k \le 2^l). \tag{12}$$

Consider the compact set $\mathscr{K} = \mathscr{K}(\varepsilon, v, p)$ constructed in the preceding lemma, and define

$$\mathscr{K}_A^{(l)} = \overline{T_l^{-1}(\mathscr{K}^l \cap I)}; \qquad \mathscr{K}_A = \bigcap_l \mathscr{K}_A^{(l)}. \tag{13}$$

(The bar denotes closure.) We have

$$\mathscr{K}^l = \bigcup_{j \in \sigma_l} \Delta_l^{(j)} \tag{14}$$

where σ_l is some subset of interval $[1, 2, \dots, 2^l]$ of natural numbers. It follows from (12) that

$$T_m^{-1}[\Delta_l^{(j)} \cap I] = T_l^{-1}[\Delta_l^{(j)} \cap I] \quad (\forall m > l),$$

whence by (14) and the containments $\mathscr{K}^1 \supset \mathscr{K}^2 \supset \cdots$ we obtain

$$\mathscr{K}_A^{(1)} \supset \mathscr{K}_A^{(2)} \supset \cdots. \tag{15}$$

Therefore

$$\mu \mathscr{K}_A = \lim_{l \to \infty} \mu \mathscr{K}_A^{(l)} = \lim_{l \to \infty} \mu T_l^{-1}(\mathscr{K}^l \cap I) = \lim_{l \to \infty} \mu(\mathscr{K}^l \cap I) = \mu \mathscr{K}$$

and $\mathscr{K}_A$ is thus a compact set of measure zero.

Suppose f satisfies the conditions of the lemma. Then we see from (13) and (15) that for some l_0,

$$|f(x)| < \varepsilon \quad (x \in \mathscr{K}_A^{(l_0)}). \tag{16}$$

Define

$$\tilde{f}(t) = f(T_{l_0}^{-1} t); \qquad F(t) = \frac{1}{|\Delta_{l_0}^{(j)}|} \int_{\Delta_{l_0}^{(j)}} \tilde{f}(\xi) d\xi \quad (t \in \Delta_{l_0}^{(j)}).$$

This defines the function F on I. From (13) and (16) it follows that

$$|\tilde{f}(t)| < \varepsilon \quad (t \in \mathscr{K}^{l_0} \cap I).$$

Because of (14), this inequality still remains after averaging; that is,

$$|F(t)| < \varepsilon \quad (t \in \mathscr{K}^{l_0} \cap I). \tag{17}$$

We obtain from (12), taking into account the properties of the mapping T_{l_0},

$$c_k^\chi(F) = \int_I F \chi_k dt = \int_I \tilde{f} \chi_k dt = \int_I f \chi_k^A dx = c_k^{\chi^A}(f) \quad (1 \le k \le 2^{l_0}); \tag{18}$$

in particular (for $k = 1$),

$$\int_I F dt = \int_I f dx. \tag{19}$$

The function F is constant on each interval $\Delta_{l_0}^{(j)} \cap I$; therefore its Fourier-Haar coefficients are different from zero only for $k \le 2^{l_0}$. Thus, in view of (18), we have

$$\|F\|_{p, \chi}^p = \sum_{k=1}^{2^{l_0}} |c_k^\chi(F)|^p = \sum_{k=1}^{2^{l_0}} |c_k^{\chi^A}(f)|^p \le \|f\|_{p, \chi^A}^p < v^p.$$

This combines with (17) and Lemma 1 to give

$$\left| \int_0^1 F \, dt \right| < 2\varepsilon,$$

which together with (19) implies inequality (5).

Lemma 3. *Suppose ϕ is a complete ONS on $[a,b]$, and an interval $\Delta \subset [a,b]$ is given. Then for any $v, \varepsilon > 0$ and $p \in [1,2)$, there exists a compact set $\mathscr{K} \subset \Delta, \mu \mathscr{K} = 0$, such that whenever $f \in L^2$, $\|f\|_{p,\phi} < v$ and $|f(x) - \alpha| < \varepsilon$ in some neighborhood of $\mathscr{K}$ (for some α, $|\alpha| < v$), the inequality*

$$\left| \frac{1}{|\Delta|} \int_\Delta f \, dx - \alpha \right| < 2\varepsilon$$

holds. (Here and in what follows, neighborhoods are understood relative to Δ.)

A linear change of variable reduces this proposition to the special case when $\Delta = [0,1]$. We extend the definition of the Haar functions (and linear combinations of them) to $[a,b]$ by setting them equal to zero outside Δ. This associates to each sequence Λ, in accordance with § 1, the system χ^Λ, orthonormal on $[a,b]$. Reasoning as we did in the proof of the theorem of § 1, we can for given $\varepsilon_k = (2^k v)^{-1}$ define $\Lambda = \{\lambda_k \uparrow\}$ and numbers $\{n_k \uparrow\}$ so as to satisfy the condition

$$\hat{\chi}_k \equiv \chi_k^\Lambda = \sum_{n=n_k+1}^{n_{k+1}} \alpha_n \phi_n + \rho_k, \qquad \alpha_n = (\hat{\chi}_k, \phi_n), \qquad \|\rho_k\|_{L^2[a,b]} < \varepsilon_k.$$

Based on Lemma 2, we choose a compact set $\mathscr{K} = \mathscr{K}_\Lambda(\varepsilon, 2v+1, p)$. Suppose the function f satisfies the conditions of this lemma. We have

$$|c_k^{\hat{\chi}}(f)| = \left| \int_\Delta f \hat{\chi}_k \, dx \right| \leq \left| \int_a^b f \sum_{n_k+1}^{n_{k+1}} \alpha_n \phi_n \, dx \right| + \|f\|_2 \varepsilon_k$$

$$\leq \sum_{n=n_k+1}^{n_{k+1}} |\alpha_n c_n^\phi(f)| + \varepsilon_k \|f\|_{p,\phi} \leq \left(\sum_{n=n_k+1}^{n_{k+1}} |c_n^\phi(f)|^p \right)^{1/p} + \frac{1}{2^k},$$

whence

$$\|f\|_{p,\hat{\chi}} = \left(\sum_k |c_k^{\hat{\chi}}(f)|^p \right)^{1/p} \leq \left(\sum_k \sum_{n_k+1}^{n_{k+1}} |c_n^\phi(f)|^p \right)^{1/p} + \left(\sum_k 2^{-kp} \right)^{1/p} \leq \|f\|_{p,\phi} + 1 < v+1.$$

By the choice of the compact set $\mathscr{K}$, this inequality together with the condition of the lemma gives inequality (5) for the function $f' = f - \alpha$, as was required.

Lemma 4. *Let $\phi, \Delta, v, \varepsilon$ and p satisfy the conditions of the preceding lemma, and let functions $f, h \in C(\Delta)$ be given. Then there exists a compact set $\mathscr{K} \subset \Delta, \mu \mathscr{K} = 0$, such that for any function $F \in L^2[a,b]$ satisfying condition (3) and*

$$\|F\|_{p,\phi} < v, \tag{20}$$

there is the inequality

$$\left| \int_\Delta (F - f) h \, dx \right| < \varepsilon. \tag{21}$$

Clearly, we may assume that $\max\{\|f\|_{C(\varDelta)}, \|h\|_{C(\varDelta)}, |\varDelta|\} < 1 < \nu$. Approximate f and h by step functions $\tilde{f}$ and $\tilde{h}$ without increase in norm, satisfying the inequality

$$\max\left\{\sup_{x\in\varDelta}|f-\tilde{f}|, \sup_{x\in\varDelta}|h-\tilde{h}|\right\} < \varepsilon' = \frac{\varepsilon}{6\nu}. \tag{22}$$

Suppose $\tilde{f}$ and $\tilde{h}$ are constant on the intervals $\{\delta_i\}$ ($1\le i\le n$, $\delta_i\cap\delta_j=\emptyset$ for $j\ne i$, $\bigcup\bar{\delta}_i=\varDelta$). Write

$$\tilde{f}\big|_{\delta_i} = \alpha_i, \qquad \tilde{h}\big|_{\delta_i} = \beta_i.$$

Using Lemma 3, for each i we can pick a compact set $\mathscr{K}_i\subset\varDelta_i=\bar{\delta}_i$ such that whenever a function F satisfies condition (20) and

$$|F(x)-\alpha_i| < \frac{\varepsilon}{4} \quad \text{in some neighborhood of } \mathscr{K}_i, \tag{23}$$

then

$$\left|\int_{\delta_i}(F(x)-\alpha_i)dx\right| < \frac{\varepsilon}{2}|\delta_i|. \tag{24}$$

Let $\mathscr{K}=\bigcup\mathscr{K}_i$. Suppose F satisfies the conditions of the lemma. Then, because of (3), condition (23) is fulfilled for all i, and consequently so is (24). From the latter it is evident that

$$\left|\int_{\varDelta}(F-\tilde{f})\tilde{h}dx\right| \le \sum_{i=1}^{n}\left|\int_{\delta_i}(F-\alpha_i)\beta_i dx\right| \le \frac{\varepsilon}{2}\sum_{i=1}^{n}|\beta_i||\delta_i| < \frac{\varepsilon}{2}.$$

Further, because of (22), we have

$$\left|\int_{\varDelta}(F-f)hdx - \int_{\varDelta}(F-\tilde{f})\tilde{h}dx\right| \le \left|\int_{\varDelta}(f-\tilde{f})\tilde{h}dx\right| + \left|\int_{\varDelta}(F-f)(h-\tilde{h})dx\right|$$

$$\le \varepsilon'+\varepsilon'+\|F\|_2\|h-\tilde{h}\|_2 \le \varepsilon'(2+\|F\|_{p,\phi}) < \frac{\varepsilon}{2}.$$

From the inequalities just obtained, (21) follows, and Lemma 4 is proved.

The Basic Lemma is deduced from this by a standard device based on the Hahn-Banach theorem. Namely, we consider the mapping

$$S: C[a,b]\to l_q \quad \left(\frac{1}{p}+\frac{1}{q}=1\right)$$

defined by the formula $Sh\equiv\hat{h}=\{c_k^\phi(h)\}$. Let $C^\varDelta\subset C[a,b]$ be the subset consisting of the functions vanishing outside $\varDelta$. The equation $\Phi(\hat{h})=\int_a^b fhdx$ defines a linear functional on the linear space $S(C^\varDelta)$, and

$$\|\Phi\| > \nu. \tag{25}$$

Otherwise we could continue Φ to a linear functional $\tilde{\Phi}$ on all of l_q with $\|\tilde{\Phi}\|\le\nu$. Let

$$\tilde{\Phi}(z) = \sum y_k z_k, \quad y=\{y_k\}, \quad \|y\|_{l_p}\le\nu \quad (\forall z=\{z_k\}\in l_q).$$

Put $\tilde{f} = \sum y_k \phi_k$ (in L^2). We have

$$\int\limits_{\Delta} \tilde{f} h dx = \int\limits_{a}^{b} \tilde{f} h dx = \sum y_k c_k^{\phi}(h) = \tilde{\Phi}(\hat{h}) = \int\limits_{a}^{b} f h dx = \int\limits_{\Delta} f h dx \quad (\forall h \in C^{\Delta});$$

which implies $f = \tilde{f}$ a.e. on Δ. Thus the function f has a continuation $\tilde{f} \in L^2[a,b]$ with $\|\tilde{f}\|_{p,\phi} = \|y\|_{l_p} \leq v < \|f\|_{p,\phi,\Delta}$, which contradicts (2). Based on (25), we pick a function $h \in C^{\Delta}$, $\|h\|_{l_q} = 1$, satisfying the inequality $\varepsilon \equiv \left| \int\limits_{\Delta} f h dx \right| - v > 0$. Applying Lemma 4, we define the compact set $\mathcal{K}(\phi, \varepsilon, v, p, \Delta)$. Suppose F satisfies condition (3). Then (4) holds. Otherwise Lemma 4 would imply inequality (21), from which we would obtain the contradiction

$$v < \left| \int\limits_{\Delta} F h dx \right| = \left| \int\limits_{a}^{b} F h dx \right| \leq \|\hat{F}\|_{l_p} \|\hat{h}\|_{l_q} = \|F\|_{p,\phi} < v.$$

Theorem 2 follows easily from the Basic Lemma. Indeed, suppose a complete ONS ϕ is fixed, and $f \in Cr^{loc}(\phi)$. Consider the point t_0 in whose neighborhoods the singularity is localized. Define $p_v = 2 - 1/v$ and $\Delta_v = [t_0 - 1/v, t_0 + 1/v] \cap [a,b]$. Then $\|f\|_{p_v, \phi, \Delta_v} = \infty$. Choose a compact set $\mathcal{K}_v \subset \Delta_v$, $\mu \mathcal{K}_v = 0$, so that for any $F \in L^2 \cap C(\Delta_v)$ satisfying the condition $F|_{\mathcal{K}_v} = f|_{\mathcal{K}_v}$, the inequality $\|F\|_{p_v, \phi} \geq v$ holds. The compact set $\mathcal{K} = \bigcup \mathcal{K}_v \cup \{t_0\}$ obviously satisfies the requirements of the theorem.

Let $A_p(\phi)$, $1 \leq p < 2$, denote the class of functions f whose norm $\|f\|_{p,\phi}$ is finite. We shall say that f locally belongs to $A_p(\phi)$ ($f \in A_p^{loc}(\phi)$) if for each point $t \in [a,b]$ there is some function $f_t \in A_p(\phi)$ agreeing with f on some neighborhood of this point.

A consequence of the basic lemma is

Theorem 2′. *If ϕ is a complete ONS, then for any $f \notin A_p^{loc}(\phi)$, $f \in C$ it is possible to exhibit a compact set $\mathcal{K}$, $\mu \mathcal{K} = 0$ such that no function $F \in C$, $F|_{\mathcal{K}} = f|_{\mathcal{K}}$ belongs to $A_p^{loc}(\phi)$.*

We shall call an ONS ϕ a *Wiener system* if the classes $A_p(\phi)$ are closed under multiplication by smooth (or polygonal) functions. The equality $A_p^{loc} = A_p$ holds for such systems. The proof is modeled after Wiener's local theorem (see [10]) with the aid of a suitable partition of unity.

The classical systems are Wiener systems. For example, for the trigonometric system τ we have

$$\|f \lambda\|_{p,\tau} = \|f \cdot \sum_k \lambda_k e^{ikx}\|_{p,\tau} \leq \sum_k |\lambda_k| \|f \cdot e^{ikx}\|_{p,\tau} = \sum_k |\lambda_k| \cdot \|f\|_{p,\tau}.$$

From Theorem 2′ it immediately follows that

Corollary. *If ϕ is a complete Wiener ONS, then for any function $f \in C$, $\sum |c_n|^p = \infty$, it is possible to exhibit a compact set $\mathcal{K}$ of measure zero such that every function $F \in C$, $F|_{\mathcal{K}} = f|_{\mathcal{K}}$, has this same property.*

For the trigonometric system τ and $p = 1$, this result follows from a theorem of M. G. Kreĭn (see [56]), $\mathcal{K}$ in this case turning out to be even countable.

However, the proof of Kreĭn's theorem is intrinsically tied to special properties of this particular case. In the general situation the compact set $\mathscr{K}$, the support of the singularity, is uncountable, as is seen from the following proposition.

Theorem 4. *Any continuous function on a countable compact set has a continuous continuation in the class $A_p(\tau)$ $(p>1)$ and in $A_1(\chi)$.*

Because of the known inclusions

$$H_1^\alpha \subset A_p(\tau), \; p > \frac{1}{\alpha} \quad (\text{Szász; see } [171] \text{ Chap. VI})$$

$$H_1^\alpha \subset A_1(\chi), \; \alpha > \frac{1}{2} \quad (\text{Ciesielski, Musielak } [22]),$$

it is enough to prove the following lemma.

Lemma 5. *Suppose $f \in C(\mathscr{K})$, where $\mathscr{K} \subset [a,b]$ is a countable compact set. Then there exists a function $F \in C[a,b]$ satisfying the conditions $F|_{\mathscr{K}} = f$; $F \in \bigcap\limits_{\alpha < 1} H_1^\alpha$.*

H_p^α will denote the Hölder classes in the space L^p. It suffices to show that for any convex modulus of continuity $\omega(\delta)$, $\omega'(0) = \infty$, there exists $F \in C[a,b]$, $F = f$ on $\mathscr{K}$, with $\omega_1(\delta, F) \leq \omega(\delta)$.*

It is easily seen that if $\omega(\delta)$ satisfies these conditions, then it is possible to pick a numerical sequence $\{\varepsilon_i > 0\}$ such that the functions

$$\alpha_i(\delta) = \min(\delta, \varepsilon_i)$$

will satisfy the inequality

$$\sum_{i=1}^\infty \alpha_i(\delta) \leq \tfrac{1}{4} \omega(\delta) \quad (0 \leq \delta \leq 1). \tag{26}$$

Suppose $\|f\|_{C(\mathscr{K})} < 1/2$. Expand the function f as a series $\sum\limits_1^\infty f_k(x)$ $(x \in \mathscr{K})$, $\|f_k\|_\infty \leq 2^{-k}$, where the f_k are step functions on $[a,b]$, continuous on $\mathscr{K}$. For each k, cover $\mathscr{K}$ with a finite system of disjoint open intervals $\delta_k^{(i)} = (\alpha_k^{(i)}, \beta_k^{(i)})$, $1 \leq i \leq i_k$, $|\delta_k^{(i)}| < \varepsilon_i$, each of which lies in the interior of an interval of constancy of the function f_k; that is,

$$f_k|_{\delta_k^{(i)}} \equiv b_k^{(i)}, \qquad |b_k^{(i)}| \leq \frac{1}{2^k}. \tag{27}$$

Set $F_k^{(i)}$ equal to zero outside $\overline{\delta_k^{(i)}}$, equal to 1 on the interval $[\alpha_k^{(i)} + \eta_k, \beta_k^{(i)} - \eta_k]$, and linear on the intervals $[\alpha_k^{(i)}, \alpha_k^{(i)} + \eta_k]$ and $[\beta_k^{(i)} - \eta_k, \beta_k^{(i)}]$, where $\eta_k > 0$ is so small that the last two intervals do not contain points of the set $\mathscr{K}$. Obviously $\sum\limits_{i=1}^{i_k} b_k^{(i)} F_k^{(i)}(x) \equiv f_k(x)$ $(x \in \mathscr{K})$, and hence, by the choice of $\{f_k\}$ and by uniform con-

---

* and then let $\omega(\delta) = \delta |\ln \delta|$. – *Trans.*

vergence, the function $F = \sum\limits_{k=1}^{\varkappa} \sum\limits_{i=1}^{i_k} b_k^{(i)} F_k^{(i)}(x)$ satisfies the conditions $F|_{\mathscr{K}} = f$, $F \in C[a,b]$.

Next we verify without difficulty the inequality $\omega_1(\delta, F_k^{(i)}) \leq 4\alpha_i(\delta)$, whence with (26) and (27) we derive

$$\omega_1(\delta, F) \leq \sum_{k=1}^{\varkappa} |b_k^{(i)}| \sum_{i=1}^{i_k} \omega_1(\delta; F_k^{(i)}) \leq \sum_{k=1}^{\varkappa} 2^{-k} \sum_{i=1}^{i_k} 4\alpha_i(\delta) \leq \omega(\delta),$$

proving the lemma.

We manifestly cannot strengthen the lemma by requiring $F \in H_1^1$, for this would mean that the function f would be of bounded variation. As a consequence of the obvious inclusion

$$C \cap H_1^\alpha \subset H_2^{\alpha/2} \tag{28}$$

we can assert that the function F contructed in the lemma belongs to $H_2^{1/2-\varepsilon}$ ($\forall \varepsilon > 0$). It would be interesting to learn whether it is possible to achieve the condition $F \in H_2^{1/2}$, or the stronger condition $F \in W_2^{1/2}$, i.e. $\sum |c_n^{(\tau)}(F)|^2 \cdot n < \infty$.

Nevertheless, the preceding results imply the following proposition: *an arbitrary continuous function f on a compact set of measure zero, generally speaking, cannot be continued to a continuous function F having even a minimal smoothness in the integral sense, that is, such that $F \in H_2^\alpha$ for some $\alpha > 0$.*

Indeed, if this were always possible, then because of the known containment

$$H_2^\alpha \subset A_p(\tau) \left(p > \frac{2}{2\alpha+1} \right)$$ (Szász's theorem), we would have a contradiction of

Theorem 3 for the trigonometric system. The result remains valid for the spaces H_1^α, because of (28).

It is very likely that the situation is similar for the classes $H_1^{\omega(\delta)}$ where $\omega(\delta)$ is an arbitrary modulus of continuity.

In connection with this it would be interesting to find a metric characterization, more sensitive than measure, of the compact sets $\mathscr{K}$ that satisfy the equality

$$[H_p^\alpha \cap C]|_{\mathscr{K}} = C(\mathscr{K}).$$

Finally, we make one more remark. As is known, a compact set $\mathscr{K}$ is called a *Helson set for the system* ϕ if $[A_1(\phi)]|_{\mathscr{K}} = C(\mathscr{K})$. One of the consequences of Theorem 3 is that for any complete ONS there exists a compact set of measure zero that is not a Helson set. For the trigonometric system, as Kreĭn's theorem shows, not even a countable compact set, generally speaking, is a Helson set. Here the arithmetic nature of the set $\mathscr{K}$ plays an essential role (see [56], which gives a detailed survey of results on Helson sets for the trigonometric system). At the same time, every countable set is a Helson set for the Haar system (Theorem 4). It would be of interest to determine whether there exists a *bounded* complete system for which every countable compact set is a Helson set.

We shall now show that there exist complete orthonormal systems for which nonlocal Carleman singularities are possible, that is, that the classes A_p and A_p^{loc} do not coincide. In diong so, we shall see that in Theorem 2' the class A_p^{loc} cannot be replaced by A_p.

Theorem 5. *There exist an ONS ϕ complete in $L^2[-\pi,\pi]$ and a function $f \in Cr(\phi)$ such that for any compact set $\mathcal{K}$, $[\mathcal{K} \neq \emptyset$, it is possible to exhibit a function $F_{\mathcal{K}} \in C \cap A_1(\phi)$ with $F_{\mathcal{K}} \equiv f$ on $\mathcal{K}$.*

By T we shall denote the linear space of all trigonometric polynomials. Let $\delta_1, \delta_2, \ldots$ be an enumeration of all the open intervals with rational endpoints contained in the interval $[-\pi, \pi]$. Fix a function $f \in C[-\pi, \pi]$, $f \notin T$. We will construct a sequence of natural numbers $\{k_s\}$ and functions $\phi_s^{(k)} \in C$ and $\{F_s\}$ satisfying the following conditions:

1. The system $\{\phi_s^{(k)}\}$ $(1 \leq k \leq k_s; s = 1, 2, \ldots)$ is orthonormal on $[-\pi, \pi]$;
2. $F_s \equiv f$ on $[\delta_s$ and $F_s \in \Phi_s \subset C$, where Φ_s is the set of all polynomials of the form $\displaystyle\sum_{i=1}^{s} \sum_{k=1}^{k_i} \alpha_i^{(k)} \phi_i^{(k)}$
3. $f \notin \Phi_s \oplus T$;
4. $\displaystyle\sum_{k=1}^{k_s-1} |(f, \phi_s^{(k)})|^{2-1/s} > 1$ $(s = 1, 2, \ldots)$.

Suppose the numbers $\{k_s\}$ and the functions $\{\phi_s^{(k)}\}$ $(1 \leq k \leq k_s)$ and $\{F_s\}$ have been constructed for all $s < \sigma$ in compliance with conditions 1–4. We shall describe the σ-th step of the induction. According to [120], in any subspace $P \subset L^2$ with finite codimension there exists an orthonormal basis consisting of trigonometric polynomials. Choose such a basis $\{t_\sigma^{(k)}\}$ $(1 \leq k < \infty)$ in the orthogonal complement of the subspace $\Phi_{\sigma-1}$. Let $f = g_\sigma + f_\sigma$, $g_\sigma \in \Phi_{\sigma-1}$, $f_\sigma \perp \Phi_{\sigma-1}$. In consequence of condition 3 for $s = \sigma - 1$, we have $\|f_\sigma\| > 0$. Pick the number k_σ so as to fulfill the conditions

$$\rho_\sigma^2 \equiv \sum_{k=1}^{k_\sigma-1} (f_\sigma, t_\sigma^{(k)})^2 > \frac{\|f_\sigma\|^2}{4}; \qquad k_\sigma > 1 + (2/\|f_\sigma\|)^{4\sigma-2}. \tag{29}$$

Clearly, in the subspace $\Phi'_\sigma = \{t_\sigma^{(k)}\}$ $(1 \leq k \leq k_\sigma - 1)$ we can pass to a new orthonormal basis $\{\phi_\sigma^{(k)}\}$ such that $(f, \phi_\sigma^{(k)}) = (f_\sigma, \phi_\sigma^{(k)}) = \rho_\sigma(k_\sigma - 1)^{-1/2}$ $(1 \leq k \leq k_\sigma - 1)$, whence because of (29) we shall have

$$\sum_{k=1}^{k_\sigma-1} |(f, \phi_\sigma^{(k)})|^{2-1/\sigma} \geq (\|f_\sigma\|/2)^{2-1/\sigma} (k_\sigma - 1)^{1/2\sigma} > 1.$$

Now since the quotient space C/T is infinite-dimensional, there exists a function $\psi_\sigma \in C[-\pi, \pi]$ with support in the interval δ_σ such that

$$\psi_\sigma \notin \{f\} \oplus \Phi_{\sigma-1} \oplus \Phi'_\sigma \oplus T, \tag{30}$$

where $\{f\}$ is the one-dimensional subspace spanned by the vector f. Define $F_\sigma = f + \psi_\sigma$. We choose $\phi_\sigma^{(k_\sigma)}$ to be the projection normal in L^2 of the vector F_σ on the orthonormal complement of the subspace $\Phi_{\sigma-1} \oplus \Phi'_\sigma$. Because of (30) and condition 3 for $s = \sigma - 1$ we shall have $f \notin \Phi_\sigma \oplus T$. Thus conditions 1–4 are preserved for $s = \sigma$.

Completing the constructed system $\{\phi_s^{(k)}\}$ $(s = 1, 2, \ldots)$, we obtain a complete ONS, which we shall call $\{\phi_n\}$, which satisfies all the conditions of the theorem. Indeed, because of condition 4 the function f has the Carleman singularity with respect to this system. At the same time, according to condition 2 there exist polynomials in the system ϕ coinciding with f outside any pre-assigned interval.

While the previous results reveal some common features inherent in all complete ONS, Theorem 5 indicates sharp differences in their behavior on these issues.

From this vantage point it is appropriate to bring in the following result (see [99]). *There exists a complete ONS $\{\phi_n\}$ consisting of piecewise continuous functions for which every continuous function $f \not\equiv 0$ has the Carleman singularity.*

The results presented above in this section were announced in [106].

Let us dwell a bit on some closely related questions.

The smoothness of functions with singularities. For the trigonometric system τ the containment

$$H^\alpha \subset A_p(\tau) \quad \left(\forall \alpha > \frac{1}{p} - \frac{1}{2}, \; 1 \leq p < 2 \right)$$

is well known (Bernšteĭn, Szász; see [10]). This result is definitive, i.e. $H^{1/p-1/2} \not\subset A_p(\tau)$. These results were extended by B. I. Golubov [42] to the Haar system. For general complete systems the question naturally can concern only the extension of the negative assertion. Here the basic method of this chapter is inapplicable, since the smoothness of a function can be ruined by a metric isomorphism.

The indicated generalization was obtained by B. S. Mityagin [78] (for $p=1$) and by S. V. Bochkarev [14] in the general case. Namely, the following proposition is true.

Theorem 6. *For any complete ONS $\{\phi_k\}$ and any $p \in [1,2)$, there exists a function $f \in H^{1/p-1/2}$ for which $\sum |c_n(f)|^p = \infty$.*

The proof [14] is based on a study of random broken lines in the class H^ω. Let $\{\theta_k^{(j)}\}$ $(1 \leq j \leq 2^k)$ denote a block of the Schauder system; that is, set $\theta_k^{(j)}(x)$ equal to zero outside the interval $\left(\dfrac{j-1}{2^k}, \dfrac{j}{2^k} \right)$, equal to 1 at the midpoint of this interval, and linear on each of its halves. After normalizing in L^2 we obtain an ON collection of functions $\{\psi_k^{(j)}(x)\}$ $(1 \leq j \leq 2^k)$. Clearly,

$$\|\psi_k^{(j)}\|_C < 3\sqrt{2^k}. \tag{31}$$

Define $\alpha_k^{(j)}(i) = (\psi_k^{(j)}, \phi_i)$. Then

$$\sum_i [\alpha_k^{(j)}(i)]^2 = 1; \qquad \sum_j [\alpha_k^{(j)}(i)]^2 \leq 1. \tag{32}$$

Consider the function

$$f_k = 2^{-k/2} \sum_{j=1}^{2^k} \pm \psi_k^{(j)}. \tag{33}$$

We shall show that for some choice of signs we have the inequality

$$\|f_k\|_{p,\phi} \equiv \left(\sum_i |c_i(f_k)|^p \right)^{1/p} \geq K_p \, 2^{k(1/p-1/2)} \quad (K_p > 0). \tag{34}$$

For this purpose we write the random broken line under consideration in the form

$$f_k^t(x) = 2^{-k/2} \sum_{j=1}^{2^k} \mathfrak{r}_j(t) \psi_k^{(j)}(x) \quad (\{\mathfrak{r}_k\} \text{ being the Rademacher system}),$$

and average with respect to t. We have

$$\int_0^1 \|f_k^t(x)\|_{p,\phi}^p \, dt = \int_0^1 \sum_i \left| \int_0^1 f_k^t(x) \phi_i(x) \, dx \right|^p dt$$

$$= 2^{-kp/2} \sum_i \int_0^1 \left| \sum_{j=1}^{2^k} \alpha_k^{(j)} \mathfrak{r}_j(t) \right|^p dt.$$

Estimating by means of the Khinchin inequality and using (32), we obtain

$$\int_0^1 \|f_k^t\|_{p,\phi}^p \, dt \geq K_p^p 2^{-kp/2} \sum_i \left(\sum_j [\alpha_k^{(j)}(i)]^2 \right)^{p/2}$$

$$\geq K_p^p 2^{-kp/2} \sum_i \sum_{j=1}^{2^k} [\alpha_k^{(j)}(i)]^2 = K_p^p 2^{k(1-p/2)},$$

whence it follows that for some $t \in I$, that is, for some sequence of signs in (33), inequality (34) is true. At the same time, there follows immediately from (31) and (33) an estimate of the modulus of continuity:

$$\omega(\delta, f_k) \leq \begin{cases} 6\delta 2^k, & 0 < \delta \leq 2^{-k} \\ 6, & \delta > 2^{-k}. \end{cases}$$

It is now easy to convince oneself that if the numbers k_v grow sufficiently rapidly, then the function

$$f = \sum 2^{-k_v(1/p-1/2)} f_{k_v}$$

satisfies the conditions of the theorem.

It clearly follows from the proof given that for any complete ONS it is possible to construct functions with the Carleman singularity that have a certain smoothness not depending on the system; for example, if $\omega(\delta)$ satisfies the condition $\omega(\delta)/\delta^\alpha \to \infty$ as $\delta \to 0$ ($\forall \alpha > 0$), then the class $H^\omega \cap \mathrm{Cr}(\phi)$ is nonempty. It is interesting to compare this with the result about absolute convergence almost everywhere (p. 68), which is of the opposite character.

We mention that the method of random Fourier series was used by Bochkarev to obtain other results similar to Theorem 6 (see [15]). For example, he proved the following proposition.

For any complete uniformly bounded ONS and any $\alpha \in (0,1]$, there is some function $f \in H^\alpha \cap V$ for which $\sum |c_n(f)|^{2/(2+\alpha)} = \infty^*$. For the trigonometric system this was proved by Szász, and the result in this case is exact.

The Paley feature. One says that an ONS $\{\phi_n\}$ has the *Paley feature* if for each sequence $\{d_n\} \notin l_2$ there exists some function $f \in C$ satisfying the condition $\sum |d_n c_n(f)| = \infty$. This property of a system obviously is stronger than the existence of a continuous function with the Carleman singularity.

---

* Recently Bochkarev showed [15'] that in this case for every $\omega(\delta)$, $\sum \omega(1/n)/n = \infty$, there is some function $f \in H^\omega \cap V$ for which $\sum |c_n(f)| = \infty$. In the trigonometric case, it gives the answer to a problem of Zygmund.

Paley proved that the trigonometric system has this feature. The result was extended by Orlicz to arbitrary bounded ONS. However, the Haar system plainly does not have the Paley feature. A. S. Makhmudov showed [71] that *an ONS ϕ has the Paley feature if and only if*

$$\inf \|\phi_n\|_1 > 0.$$

Makhmudov investigated in detail the properties of Fourier coefficients for such systems and extended to them the known results of S. B. Stechkin [126] concerning the trigonometric system.

Adjustment of functions. Allied with problems of continuation of functions (Theorems 3, 4) are problems of adjusting functions in order to improve the properties of the Fourier series. In contrast to the formulation of the question above, the set on which the adjustment is made is not prescribed in advance.

The point of departure in these problems is Menshov's theorem (see [10]) that *any measurable (finite) function can be changed on a set of arbitrarily small measure so that the resulting function expands as a uniformly convergent trigonometric Fourier series.*

There arises the question of whether a function can be adjusted in such a way as to make its Fourier coefficients decrease sufficiently rapidly, for example, $\{c_n\} \in l_p$ for some $p < 2$; in other words, is it possible by changing a given function on a set of small measure to remove the Carleman singularity? A negative answer seems more likely, at least for $p = 1$, but this has not been proved.

For the Haar system, as usual, the situation turns out to be simpler. Yu. S. Fridlyand showed [34'] that there exists $f \in C$ such that every function $\tilde{f}$ agreeing with f on any set of positive measure satisfies the condition $\sum |c_n^\chi(\tilde{f})|^p = \infty$ ($\forall p < 2$). It is interesting to compare this with F. G. Arutyunyan's result [4] that any measurable function can be adjusted on a set of arbitrarily small measure in such a way that the Fourier-Haar series will converge absolutely almost everywhere.

However, A. A. Talalyan showed [135] that for some complete systems the problem indicated above is solved in the affirmative: there exists a complete ONS such that for any measurable function it is possible by means of an adjustment on a set of arbitrarily small measure to achieve the condition $\sum |c_n| < \infty$. Using the method of Chap. IV in this construction, one can construct such a system from uniformly bounded continuous functions.

We take note, finally, of the following problem, which apparently is unsolved: does every continuous function on a compact set of measure zero have a continuation whose trigonometric series converges uniformly? One can show that for countable sets it is true.

§ 5. Some More Results about the Haar System

The role of the Haar system in the theory of general orthogonal systems is not limited to the foregoing. Because of the peculiarities of its construction, it is more accessible to investigation than the other classical systems, and at the same time it differs markedly from them in many respects. For these reasons the Haar system not infrequently turns up as the example in which the possibility of a given feature for an ONS is first discovered. This refers in particular to the classical results of Haar and Marcinkiewicz.

This section contains some new facts in this direction.

It should be noted that in recent years the Haar system has been subjected to thorough and detailed study. For a survey of the works available here see [43]. In what follows we shall concern ourselves only with some of the results about the Haar system that throw light on questions of the general theory.

Absolute and unconditional convergence. Recall that unconditional convergence of a series almost everywhere means convergence almost everywhere under any ordering of the terms, with the exceptional set of measure zero depending on the arrangement, so that the given series might not have any point of absolute convergence.

If $\{\phi_n\}$ is a uniformly bounded ONS, then a necessary and sufficient condition for absolute convergence almost everywhere of the series $\sum c_n \phi_n$ is $\sum |c_n| < \infty$ (see [1]). But unconditional convergence of a series with respect to any ONS is ensured by a weaker condition due to Menshov: $\{c_n\} \in \bigcup_{p<2} l_p$.

There arises the question of whether there exists a complete ONS for which unconditional convergence coincides with absolute convergence. As it turns out, the Haar system has this property. The following theorem is true, proved by E. M. Nikishin and P. L. Ulyanov [88].

Theorem 1. *If the series*

$$\sum a_n \chi_n(x) \tag{1}$$

unconditionally converges almost everywhere on a set E, then it converges absolutely almost everywhere on this set.

Suppose not, i. e., $\sum |a_n \chi_n(x)| = \infty$ $(x \in E' \subset E, \mu E' > 0, a_n > 0)$. Fix a sequence of indices $\{n_k \uparrow\}$ for which

$$\sum_{n=n_k+1}^{n_{k+1}} |a_n \chi_n(x)| > k \quad \left(x \in E_k \subset E'; \mu(E' \backslash E_k) < \frac{1}{k}; k = 1, 2, \dots\right).$$

Making the ω-rearrangement of each of the segments of the series $\sum_{n=n_k+1}^{n_{k+1}} a_n \chi_n$ (§ 2), we obtain a series diverging almost everywhere on E', because of inequality (4) of § 2.

We mention that for monotonic coefficients Theorem 1 was established by Ulyanov in [155]. See [140] for some generalizations.

From Theorem 1 follows a characterization of Weyl multipliers for the unconditional convergence of Haar series, found by P. L. Ulyanov [156] (see also F. Móricz [79']).

Theorem 2. *A necessary and sufficient condition for a sequence $\omega(n)\uparrow$ to be a Weyl multiplier for unconditional convergence of Haar series is the condition*

$$\sum \frac{1}{n\omega(n)} < \infty. \tag{2}$$

Indeed, suppose (2) is fulfilled and $\sum c_n^2 \omega(n) < \infty$. Then we have

$$\sum |c_n|\, \|\chi_n\|_1 \leq \sqrt{2}\sum |c_n|/n^{1/2} = \sqrt{2}\sum |c_n|\sqrt{\omega(n)}\cdot\frac{1}{\sqrt{n\omega(n)}} < \infty,$$

whence $\sum |c_n \chi_n(x)| < \infty$ almost everywhere.

Now suppose the sequence $\omega(n)\uparrow$ does not satisfy (2); that is, $\sum \dfrac{1}{\omega(2^n)} = \infty$. Then we can pick a sequence $b_k \geq 0$ so that

$$\sum b_k = \infty, \qquad \sum b_k^2 \omega(2^{k+1}) < \infty. \tag{3}$$

For this it suffices to choose numbers k_i so that $\sum \dfrac{1}{D_i} < \infty$, where $D_i = \displaystyle\sum_{k=k_i+1}^{k_{i+1}} \frac{1}{\omega(2^{k+1})}$, and to set $b_k = \dfrac{1}{D_i\omega(2^{k+1})}$ $(k_i < k \leq k_{i+1})$. Then the Haar series

$$\sum_{k=1}^{\infty} b_k 2^{-k/2} \sum_{n=2^k+1}^{2^{k+1}} \chi_n \equiv \sum a_n \chi_n$$

satisfies the conditions

$$\sum a_n^2 \omega(n) \leq \sum b_k^2 \omega(2^{k+1}) < \infty; \qquad \sum a_n |\chi_n(x)| = \sum b_k = \infty,$$

which together with Theorem 1 imply that $\omega(n)$ is not a Weyl multiplier for unconditional convergence.

This result shows that the Haar system does not have an exact Weyl multiplier for unconditional convergence. It is not known how matters stand on this issue for other complete systems.

For the trigonometric system we do not know any sufficient conditions for Weyl multipliers for unconditional convergence that are stronger than Orlicz's condition

$$\omega(n)\ln^{-2}n\uparrow, \qquad \sum \frac{\ln n}{n\omega(n)} < \infty, \tag{4}$$

which is valid for any ONS (see [158]).

A lower bound in this case has been established by Móricz [79]: if $\omega(n)$ is a Weyl multiplier for unconditional convergence of trigonometric series, then $\omega(n) \neq o(\ln\ln n)$. This theorem was preceded by an analogous result for the Walsh system, established by Tandori [143]. Nakata [82], making Móricz's construction more precise, obtained a stronger lower bound for Weyl multipliers for the trigonometric system: $\omega(n) \neq o(\sqrt{\ln n})$. A discrepancy between this estimate and condition (4) still remains: $(\omega(n)=\ln^\alpha n, \frac{1}{2} \leq \alpha \leq 2)$.

For the Haar system there have also been investigations of smoothness conditions on a function that ensure unconditional (hence absolute) convergence almost everywhere of its Fourier series. Specifically, Ulyanov showed [156] that if $f \in H^\omega$, $\omega(\delta) = (-\ln \delta)^{-(1/2 + \varepsilon)}$, $\varepsilon > 0$, then the Fourier-Haar series of this function converges absolutely almost everywhere. At the same time, Bochkarev [13], developing the construction of [93], constructed an example of a function f with $\omega(\delta, f) = O(-\ln \delta)^{-1/2}$ for which $\sum |c_n \chi(x)| = \infty$ almost everywhere.

For unconditional convergence of trigonometric Fourier series, the corresponding question remains open.

The representation problem. A fundamental theorem of Menshov says that any finite measurable function can be represented as a trigonometric series that converges to it almost everywhere.

The analogous result for the Haar system was established by N. K. Bary and strengthened by F. G. Arutyunyan [4]: every finite measurable function can be expanded as a Haar series that converges absolutely almost everywhere.

The problem of whether a trigonometric series can tend to $+\infty$ on a set of positive measure is unsolved. It has been discovered that for the Haar system this problem has a negative solution. Namely, A. A. Talalyan and F. G. Arutyunyan proved the following proposition [139].

Theorem 3. *Series* (1) *cannot tend to* $+\infty$ *on a set of positive measure.*

A beautiful proof of this theorem was given by V. A. Skvortsov [122]. That author noticed an interesting connection between the convergence of series (1) and the concept of differentiability with respect to a net (see [115]).

We shall call the set of points $j/2^k$ $(0 \le j \le 2^k)$ the elements of the net R_k. Suppose a function ψ is defined on the set R of dyadic-rational points of the interval $[0, 1]$. Assign to each point $x \in I = \complement R$ and to each k the pair of points $a_k = j/2^k$ and $b_k = (j + 1)/2^k$ such that $a_k < x < b_k$. The derivative of the function ψ at the point x with respect to the sequence of nets $\{R_k\}$ will mean the limit

$$RD\psi(x) = \lim_{k \to \infty} \frac{\psi(b_k) - \psi(a_k)}{b_k - a_k}$$

(if it exists).

Now associate to any series $\sum_1^\infty c_k \chi_k(x)$ the function

$$\psi(x) = \sum_{k=1}^\infty c_k \int_0^x \chi_k(t) dt,$$

defined on the set R. It is easy to see that convergence of the series at a point $x \in I$ is equivalent to RD-differentiability of the function ψ at this point, with the sum of the series being $S(x) = RD\psi(x)$.

It is well known that a derivative (in the ordinary sense) cannot equal $+\infty$ on a set of positive measure. An analogous fact is also true for differentiation with respect to nets (see [115]). From this follows the assertion of the theorem.

This result shows that the so-called representation problem in the broad sense (when every measurable function—not necessarily finite— is required to be represented as an almost everywhere convergent series with respect to a given system $\{\phi_k\}$) can have a negative solution in the class of complete systems.

Curiously enough, Theorem 3 is intrinsically tied to the usual ordering of the Haar system. That is, the following proposition is true [102].

Theorem 4. *There exists a rearrangement $\{\chi_{n_k}\}$ of the Haar system for which the representation problem in the broad sense has an affirmative solution.*

The idea of the proof is as follows: it is not difficult to prove that all the partial sums of the polynomial

$$\sum_{k=1}^{n} -2^{\frac{n-k}{2}} \chi_{n-k}^{(1)}(x) \tag{5}$$

are ≥ 1 outside the interval $\Delta_n^{(1)}$. On each interval of constancy of this polynomial it is possible to concentrate a new Haar polynomial of type (5) possessing a similar property: all of its partial sums will be ≥ 1 outside an interval of arbitrarily small length. Continuing this process infinitely many times and choosing the orders of the polynomials so that the sum of the measures of the exceptional sets will be finite, we obtain a series with respect to the rearranged Haar system $\{\chi_{n_k}\}$ that tends to $+\infty$ almost everywhere. It is easily seen that for any measurable set $E \subset [0,1]$ one can select from this series a subseries tending to $+\infty$ almost everywhere on E and having a finite sum almost everywhere outside E; for this it suffices to remove from the series all the terms on whose supports the set E has density $<1/2$. Recalling further Arutyunyan's generalization of Bary's theorem (see above), we obtain the conclusion of the theorem. It is not out of the question that such a result might hold for any complete system.

It is to be noted that while the representation problem in the broad sense is answered generally in the negative, the representation problem in the narrow sense (when the concern is for the representation of finite functions by series convergent almost everywhere) is in general unsolved. Here we know the following result, due to R. Davtyan: *if $\{\phi_n\}$ is a complete ON system of convergence and satisfies the localization principle, then every measurable finite function is representable as an almost everywhere convergent series with respect to this system.*

This result means in particular that Menshov's theorem for the trigonometric system is a consequence of Carleson's theorem and the Riemann localization principle.

We note also the following proposition (Arutyunyan [4′]): *for any basis in the space C, the answer to the restricted representation problem is affirmative.*

At the same time, for general complete ONS it is apparently not even known whether there can always be found an almost everywhere convergent series $\sum c_n \phi_n$ with coefficients $\{c_n\} \notin l_2$.

For the case of convergence in measure, the answer to the representation question is affirmative for any complete ONS, as shown by A. A. Talalyan (see [134]). For a detailed survey of works in this area see [138], [160].

Convergence almost everywhere. We have seen (§ 3) that in the study of convergence in mean of series (1), an important role is played by the quantity

$$\sum c_n^2 \chi_n^2(x). \tag{6}$$

This same quantity regulates convergence of series (1) almost everywhere. The following proposition has been proved by Arutyunyan [4] and by Gundy [44] (the latter proved it for general martingales): *For the convergence of series (1) almost everywhere on a set E, it is necessary and sufficient that the sum (6) be finite almost everywhere on E.* Hence it follows that convergence almost everywhere of series (1) implies convergence almost everywhere of any subseries. This is clearly not the case for the trigonometric system.

It follows in particular from what has been said that *every Fourier-Haar series converges unconditionally in measure* (as unconditional convergence in a complete linear metric space is equivalent to convergence of all the subseries (Orlicz; see [55])).

In the case of monotone coefficients, the finiteness of sum (6) on a set of positive measure is equivalent, as is easily verified, to the condition $\sum c_n^2 < \infty$. Hence we obtain Ulyanov's result [155] that *series (1) with $c_n \downarrow 0$ converges almost everywhere (or on E, $\mu E > 0$) if and only if $\sum c_n^2 < \infty$.*

It is not known whether this condition can be the criterion for convergence almost everywhere (without any additional conditions) for some complete ONS.

Approximation properties of Fourier series. B. Szőkefalvi-Nagy [132] gave the following estimate for the order of approximation of continuous functions by the partial sums of their Fourier-Haar series:

$$\sup_t |f(t) - S_n(t)| \le K\omega\left(\frac{1}{n}; f\right) \tag{7}$$

(about the value K see [43]). P. L. Ulyanov [157] extended inequality (7) to the L^p-spaces for $1 \le p < \infty$.

Relation (7) is yet another manifestation of the extremal properties of the Haar system. As Sz.-Nagy showed (see [1]), the estimate

$$\|f - S_n\|_C = o\left(\omega\left(\frac{1}{n}; f\right)\right) \tag{8}$$

is impossible for any ONS $\{\phi_n\}$ if o is to be uniform on the class C.

This problem admits of a geometric interpretation relating it to Kolmogorov's theory of diameters. In particular, it follows from the results of [149] that relation (8) cannot hold uniformly on any compact set $H^\omega \subset C$. Something of a supplement to this is contained in [96], where it is shown that relation (8) cannot hold for any system even with o depending on f.

The order of growth of non-oscillating orthonormal sequences. In this subsection we present one more result characterizing the extremal properties of the Haar system.

Suppose the interval $[a,b]$ can be divided into a finite number of intervals on each of which a given function f is of constant sign (i.e., one of the two inequalities $f(x) \ge 0$, $f(x) \le 0$ holds almost everywhere). The minimal number of pieces for such a partition will be called the number of changes of sign of the function f on $[a,b]$, and will be denoted by $Z[f]$.

Haar proved [49] that if $\{\phi_k\}$ $(1 \le k \le n)$, $|\phi_k| \le M$, is a collection of functions orthonormal on $[a,b]$, then the inequality

$$\max_{1 \le k \le n} Z[\phi_k] \ge K \frac{n}{M^4} \tag{9}$$

holds, where $K > 0$ is an absolute constant. Hence, in particular, it follows that if $\{\phi_k\}$ is an ONS and

$$\|\phi_n(x)\|_\infty \equiv M_n = o(n^{1/4}), \tag{10}$$

then there follows the condition

$$\limsup_{n \to \infty} Z[\phi_n] = \infty. \tag{11}$$

On the other hand, we see from the example of the Haar system that growth of the order of $M_n = O(n^{1/2})$ is compatible with the condition $Z[\phi_n] < K$. The question arises: what is the maximal order of growth of a system that still makes oscillation (that is, condition (11)) inevitable?

Krantsberg showed [63] that this condition is always fulfilled as soon as

$$M_n = O(n^{1/2 - \varepsilon}) \quad \text{(for some } \varepsilon > 0).$$

This sharpens estimate (10) as far as is possible in the power scale. We indicate below a simple device which gives a more exact result.

Theorem 5. *If*

$$M_n = o(n^{1/2}/\ln n),$$

then condition (11) holds.

The basic role here is played by the following lemma, which gives an upper bound for the primitives of any orthonormal sequence of functions.

Lemma. *Let $\{\phi_n\}$ be an ONS on $[a,b]$ and let $\Phi_n(x) = \int_a^x \phi_n(t)dt$. Then the following inequality is true for infinitely many indices n:*

$$\|\Phi_n\|_{C[a,b]} < K \frac{\ln n}{n^{1/2}}.$$

Let $[a,b] = [0,\pi]$. We have that $\phi_n \overset{(L^2)}{=\!=\!=} \sum_{k=0}^{\infty} a_k^{(n)} \cos kt$, whence

$$\Phi_n(x) = a_0^{(n)} x + \sum_{k=1}^{\infty} \frac{a_k^{(n)}}{k} \sin kt; \qquad \|\Phi_n\|_{C[0,\pi]} \le \pi|a_0^{(n)}| + \sum_{k=1}^{\infty} |a_k^{(n)}|/k \le K_1 \sum_{k=0}^{\infty} \frac{|a_k^{(n)}|}{k+1}.$$

Further,

$$\sum_{n=1}^{N} \|\Phi_n\| \le K_1 \left[\sum_{n=1}^{N} \sum_{k=0}^{N} |a_k^{(n)}|/(k+1) + \sum_{n=1}^{N} \sum_{k=N+1}^{\infty} |a_k^{(n)}|/(k+1) \right]$$

$$\le K_1 \left[\sum_{k=0}^{N} \frac{1}{k+1} \sum_{n=1}^{N} |a_k^{(n)}| + \sum_{n=1}^{N} \left(\sum_{k=N+1}^{\infty} |a_k^{(n)}|^2 \right)^{1/2} \left(\sum_{N+1}^{\infty} (k+1)^{-2} \right)^{1/2} \right]$$

$$\le K_2[\sqrt{N} \ln N + \sqrt{N}] < K_3 \sqrt{N} \ln N,$$

from which the conclusion of the lemma follows.

Considering the inequalities

$$1 = \int_a^b \phi_n^2(x)\,dx \le M_n \int_a^b |\phi_n(x)|\,dx = M_n \sum_{k=1}^{Z[\phi_n]} \left| \int_{\delta_k^{(n)}} \phi_n(x)\,dx \right| \le 2M_n Z[\phi_n] \|\Phi\|_C$$

(where $\{\delta_k^{(n)}\}$ is the partition of the interval $[a,b]$ into the $Z[\phi_n]$ intervals of constancy of sign of the function ϕ_n), we convince ourselves of the truth of the theorem.

We do not know whether it is possible to dispense with the logarithmic factor in Theorem 5. It would also be interesting to make precise the relationship between the orders of growth of the numbers M_n and $Z[\phi_n]$; in particular, might it be possible to put $K\dfrac{n}{M^2}$ in the right-hand side of inequality (9)?

Chapter IV. Series from L^2 and Peculiarities of Fourier Series from the Spaces L^p

This chapter is devoted to the convergence almost everywhere and in mean of Fourier series with respect to general orthogonal systems. However, in contrast to Chap. II, properties here are first stipulated for the function, and not for the coefficients of the expansion.

Considerable attention has been given to the peculiarities of Fourier series in the spaces L^p, $p<2$. In this case, effects arise that differ sharply from the properties of L^2-series. For example, it turns out (§ 4) that a Fourier series with respect to a complete system, even if the series converges almost everywhere, does not always converge to the original function.

In §§ 2–3 we investigate the role of the Lebesgue functions in convergence and summability almost everywhere, and the connection between the convergence of Fourier series of functions from various classes. The initial result in this area of examination is the fundamental Kolmogorov-Seliverstov-Plessner theorem.

To a considerable extent, the contents of this chapter introduce constructions of orthonormal systems that reveal the possibility of "nonclassical" properties. An essential role in this is played by the special finite-dimensional transformations given by the matrices A_k, which were introduced in the work [98]. Given an ONS with certain properties, features of the structure of these matrices permit us to construct in some cases (§§ 3–4) a uniformly bounded complete ONS with these same properties.

The matrices A_k find application also in constructions in another scheme (§§ 2, 5). In particular, they permit us to accomplish a geometric construction of conditional bases in Hilbert space, which to a certain extent casts light on the nature of such bases.

The contents of §§ 2, 3 of this chapter are connected with the results of § 2, Chap. III. The ω-rearrangements of the Haar system that were introduced there serve here as a useful tool.

§ 1. The Matrices A_k

Let k be an arbitrary fixed natural number. Define a square matrix $A_k=\|a_{ij}^{(k)}\|$ in the following way:

$$a_{ij}^{(k)} = \frac{1}{\sqrt{2^k}} \chi_j(t_i) \quad (1\le i,j\le 2^k), \tag{1}$$

where $\{\chi_j\}$ is the Haar system, and t_i are points lying respectively in the intervals $\left(\dfrac{i-1}{2^k}, \dfrac{i}{2^k}\right)$.

The colums of this matrix represent a basis of the Haar type in the Euclidean space $\mathbb{R}^{2^k}$. The following properties are obvious.

(i) *The matrix A_k is orthogonal.*

This fact follows immediately from the orthogonality of the Haar system:

$$\sum_{i=1}^{2^k} a_{ij}^{(k)} a_{ir}^{(k)} = \frac{1}{2^k} \sum_{i=1}^{2^k} \chi_j(t_i)\chi_r(t_i) = \int_0^1 \chi_j(t)\chi_r(t)\,dt.$$

(ii) *The inequality*

$$\sum_j |a_{ij}^{(k)}| < C \quad (\forall i, k)$$

holds (where C is an absolute constant).

It is sufficient to notice that for each i exactly one function from the s-th block of the Haar system differs from zero at the point t_i and is equal in modulus to $2^{s/2}$. Therefore,

$$\frac{1}{\sqrt{2^k}} \sum_j |\chi_j(t_i)| = \frac{1}{\sqrt{2^k}}\left[1 + \sum_{s=0}^{k-1} \sqrt{2^s}\right] < C.$$

We introduce the "Dirichlet kernel" for the orthonormal system consisting of the rows of the matrix A_k:

$$d_v(j,r) = \sum_{i=1}^{v} a_{ij}^{(k)} a_{ir}^{(k)}. \tag{2}$$

(iii) *The inequality*

$$\sum_{r=2^\sigma+1}^{2^{\sigma+1}} |d_v(j,r)| \le 2^{-|\sigma-s|/2} \quad (1 \le j, v \le 2^k; 0 \le \sigma \le k-1) \tag{3}$$

is true, where $s=s(j)$ is determined by the condition $2^s < j \le 2^{s+1}$ $(s(1)=0)$.

Let j, σ, v be fixed. We have

$$d_v(j,r) = \frac{1}{2^k} \sum_{i=1}^{v} \chi_j(t_i)\chi_r(t_i) = \int_0^{v/2^k} \chi_j(t)\chi_r(t)\,dt. \tag{4}$$

The last integral can differ from zero only in the case where the point $\dfrac{v}{2^k}$ lies inside the support of the function χ_r. Clearly, this can occur for not more than one Haar function of the block $\{\chi_r\}$ $(2^\sigma < r \le 2^{\sigma+1})$. Thus, in the left side of inequality (3) only one term can differ from zero (with $r = \rho\,(j, \sigma, v)$). With this, because of (4), we have

$$|d_v(j,\rho)| \le \|\chi_{\max(j,\rho)}\|_1 \, \|\chi_{\min(j,\rho)}\|_\infty = \frac{\sqrt{2^{\min(s,\sigma)}}}{\sqrt{2^{\max(s,\sigma)}}} = \frac{1}{\sqrt{2^{|s-\sigma|}}},$$

whence (3) follows.

(iv) *Let* $\dfrac{1}{\sqrt{2}} < \alpha < \sqrt{2}$, $\lambda_j \equiv \lambda_j(\alpha, k) = \alpha^{k - s(j)}$. *Then*

$$\sum_{r=1}^{2^k} |d_\nu(j,r)| \lambda_r < C(\alpha) \lambda_j \quad (1 \leq j, \nu \leq 2^k; k = 1, 2, \ldots).$$

Indeed, taking into account the last inequality, we have

$$\sum_{r=1}^{2^k} |d_\nu(j,r)| \lambda_r = |d_\nu(j,1)| \lambda_1 + \sum_{\sigma=0}^{k-1} \sum_{r=2^\sigma+1}^{2^{\sigma+1}} |d_\nu(j,r)| \lambda_r$$

$$\leq 2^{-\frac{s(j)}{2}} \alpha^k + \sum_{\sigma=0}^{k-1} 2^{-|s(j)-\sigma|/2} \alpha^{k-\sigma}$$

$$= \lambda_j \left(2^{-\frac{s(j)}{2}} \alpha^{s(j)} + \sum_{\sigma=0}^{s(j)} \left(\frac{\alpha}{\sqrt{2}} \right)^{s(j)-\sigma} + \sum_{\sigma=s(j)+1}^{k-1} (\alpha\sqrt{2})^{s(j)-\sigma} \right)$$

$$\leq \lambda_j \left(1 + \sum_{\mu=0}^{\infty} \left(\frac{\alpha}{\sqrt{2}} \right)^\mu + \sum_{\mu=0}^{\infty} \left(\frac{1}{\alpha\sqrt{2}} \right)^\mu \right) = C(\alpha) \lambda_j.$$

The statement just proved, for $\alpha = 1$, gives an estimate of the "Lebesgue functions" for the system of rows of the matrix A_k:

(v) $$l_\nu(j) \equiv \sum_{r=1}^{2^k} |d_\nu(j,r)| < C.$$

This is an estimate, uniform with respect to k and ν, of the norm of the projections onto the subspaces generated by the row vectors in the n-dimensional, $n = 2^k$, space $l_\gamma^{(n)}$ ($\|x\| = \max_i |x^{(i)}|$). For the expansion with respect to the columns of the matrices A_k, such an estimate follows easily from the properties of the Haar system. However, for general orthogonal matrices the boundedness of the projections corresponding to the expansion relative to the columns does not imply an analogous property for the expansion relative to rows. Here we encounter an interesting feature of the matrices (1).

§ 2. Lebesgue Functions and Convergence Almost Everywhere

As we know (see Chap. I), the Lebesgue functions

$$L_n(x) = \int_a^b \left| \sum_{k=1}^n \phi_k(x) \phi_k(t) \right| dt \tag{1}$$

of an orthonormal system $\{\phi_k\}$ play a determining role in the questions of uniform (and local) convergence of the Fourier series of continuous functions, and also convergence in mean.

If the relation

$$L_n(x) < K \quad (x \in [a,b], n = 1, 2, \ldots) \tag{2}$$

is satisfied, then (with the assumption that the system is closed) this system forms a basis in the spaces C and L (and, because of the Riesz interpolation theorem, in all L^p, $1 \leq p < \infty$).

There is a connection between the behavior of the Lebesgue functions and convergence almost everywhere. This connection was discovered by Kolmogorov, Seliverstov and Plessner (for the case of the trigonometric system), and extended to the general case by Kaczmarz. The basic step is the following proposition (for the proof see [55]):

If the condition

$$L_k(x) < \omega(k) \quad (x \in E, k = 1, 2, \ldots; \omega(k) \uparrow) \tag{3}$$

is satisfied, then the inequality

$$\left\{ \int_E \max_{1 \leq k \leq n} \frac{1}{\sqrt{\omega(k)}} \left| \sum_{i=1}^{k} c_i \phi_i(x) \right| dx \right\}^2 \leq 4 \sum_{1}^{n} c_k^2 \mu E \quad (\forall \{c_k\}) \tag{4}$$

is true.

Hence it follows that for $f \in L^2$ the partial sums of its Fourier series have the order

$$S_n(x) = O_x(\sqrt{\omega(n)}) \quad \text{a.e. on } E. \tag{5}$$

An immediate consequence is

Theorem 1. *If* $L_n(x) = O_x(1)$ $(x \in E)$, *then for any function* $f \in L^2$ *the Fourier series*

$$f \sim \sum c_n \phi_n \tag{6}$$

converges almost everywhere on E.

Only technical complications arise for the more general result (see [55]): *if (3) is satisfied, then the sequence* $\omega(k)$ *is a Weyl multiplier for convergence almost everywhere on* E.

Thus, condition (2) ensures that (for closed systems) every function $f \in L^p$, $1 \leq p < \infty$ expands as a Fourier series converging to it in the L^p-metric, and for $p \geq 2$, also converging almost everywhere.

The question arises as to whether there is convergence almost everywhere in the case $p < 2$. For classical systems satisfying condition (2) this is precisely the case. In the general case, however, the answer turns out to be negative [98].

Theorem 2. *There exists an ONS* $\{\phi_n\}$, *closed in C and satisfying condition (2), such that for some function* $f \in \bigcap_{p < 2} L^p$ *the Fourier series (6) diverges almost everywhere.*

We apply the matrix A_k (§ 1) to the k-th Haar block:

$$\phi_i^{(k)} = \sum_{j=1}^{2^k} a_{ij}^{(k)} \chi_k^{(j)} \quad (1 \leq i \leq 2^k, k \geq 1).$$

The system $\phi = \{\chi_1, \chi_2, \bigcup_k \{\phi_i^{(k)}\}\}$ is orthonormal and closed in $C[0,1]$ by property (i). With this,

$$\sum_{i=1}^{v} \phi_i^{(k)}(x)\phi_i^{(k)}(t) = \sum_{i=1}^{v}\sum_{r=1}^{2^k}\sum_{j=1}^{2^k} a_{ij}^{(k)} a_{ir}^{(k)} \chi_k^{(j)}(x) \chi_k^{(r)}(t)$$

$$= \sum_{r}\sum_{j} d_v(j,r) \chi_k^{(j)}(x) \chi_k^{(r)}(t),$$

where d_v is defined by relation (2) § 1. Further,

$$L_v^{(k)}(x) \equiv \int_0^1 \left| \sum_{i=1}^{v} \phi_i^{(k)}(x)\phi_i^{(k)}(t) \right| dt \le \int_0^1 \sum_{j} \left| \sum_{r} d_v(j,r) \chi_k^{(r)}(t) \right| |\chi_k^{(j)}(x)| dt$$

$$\le \max_j \left[\sqrt{2^k} \int_0^1 \left| \sum_{r} d_v(j,r) \chi_k^{(r)}(t) \right| dt \right]$$

$$\le \max_j \left[\sqrt{2^k} \sum_{r} |d_v(j,r)| \, \|\chi_k^{(r)}\|_1 \right] = \max_j l_v(j) < C$$

(the last inequality is because of property (v) § 1). In view of Lemma 1 § 2, Chap. I, we conclude that the system ϕ satisfies relation (2).

Suppose that $f = \sum_{k=1}^{\infty} k \chi_k^{(1)}$ (the series converges in L^p, $p<2$). Since $\chi_k^{(1)} = \sum \dfrac{1}{\sqrt{2^k}} \phi_i^{(k)}$, the Fourier series (6) can be written in the form

$$\sum_{k}\sum_{i=1}^{2^k} \frac{k}{\sqrt{2^k}} \phi_i^{(k)}. \tag{7}$$

Let x be a dyadic-irrational point, $\dfrac{1}{2^m} < x < \dfrac{1}{2^{m-1}}$, $m \ge 1$. Fix $k>m$, and define a number $j_0=j(k,x)$ by the condition that $\dfrac{j_0-1}{2^k} < x < \dfrac{j_0}{2^k}$. Clearly, $2^{k-m}<j_0 \le 2^{k-m+1}$. We have, because of (1) § 1,

$$\max_{1 \le i \le 2^k} |\phi_i^{(k)}(x)| = \max_i \left| \sum_{j} a_{ij}^{(k)} \chi_k^{(j)}(x) \right| = \max_i |a_{ij_0}^{(k)}| \sqrt{2^k}$$

$$= \max_i |\chi_{j_0}(t_i)| = \|\chi_{j_0}\|_\infty = \sqrt{2^{k-m}} .$$

Now letting $k \to \infty$, we conclude that the general term of series (7) is not bounded at the point x; that is, the series (6) diverges almost everywhere, and the theorem is proved.

The series constructed clearly diverges almost everywhere for any order of the terms. At the same time, the system ϕ is a basis in every L^p. Hence it follows that *a series with respect to orthogonal functions can converge in mean (in the L^p-metric, $p<2$) and not converge almost everywhere for any ordering of the terms.*

Thus, the theorem of Garsia (Chap. II, § 2) does not extend to the spaces L^p, $p<2$. This reveals yet another distinction between Fourier series in these spaces and those in the space L^2.

We mention a related result [98]: *there exist a complete uniformly bounded ONS $\{\phi_n\}$ and a function $f \in \bigcap_{p<2} L^p$ whose Fourier series (6) diverges almost everywhere, and even in measure, for any ordering of the terms.*

The impossibility of improving Theorem 1, from the point of view of the coefficients, was shown by Tandori [146]: *for any sequence $\{\alpha_n\} \notin l_2$ there exists an ONS ϕ, with condition (2), for which the series $\sum \alpha_n \phi_n$ diverges almost everywhere.* (In the case $\alpha_n \downarrow$, ϕ can be taken to be the Haar system, according to a result of Ulyanov; see Chap. III, § 5.) In the theorem stated, however, the series $\sum \alpha_n \phi_n$ is not in general a Fourier series. It is probable that this theorem can be combined with Theorem 2.

Lebesgue functions and summability of Fourier series. A natural area for the generalization of Theorem 1 and the results connected with it is the summability of orthogonal series. If $T = \|\alpha_{nj}\|$ $(1 \le n, j < \infty)$ is a Toeplitz matrix determining a regular method of summation for numerical sequences (see [171]), $\alpha_{nj} = 0$ $(j > j_n)$, then the T-means of the Fourier series (6) can be written in the form

$$\sigma_n(f; x) = \sum_{j=1}^{j_n} \alpha_{nj} S_j(f; x) = \int_a^b f(t) \mathscr{D}_n(T, x, t)\, dt,$$

$$\mathscr{D}_n = \sum_{j=1}^{j_n} \alpha_{nj} \sum_{l=1}^{j} \phi_l(x)\, \phi_l(t).$$

The role of the Lebesgue functions is played by the quantities $L_n(T, x)$ $= \int_a^b |\mathscr{D}_n(T, x, t)|\, dt$. The results on local and uniform convergence, and also convergence in the L^p-metric, that were referred to at the beginning of this section, and that depend on general theorems on linear operators, extend to this case (the Toeplitz conditions ensure the convergence of the operators σ_n on the set of polynomials $f = \sum_1^{k(f)} a_k \phi_k$, and the condition

$$L_n(T, x) < K \quad (x \in [a, b], n = 1, 2, \ldots) \tag{8}$$

gives the uniform boundedness of these operators).

To some extent results on convergence almost everywhere also still hold. Kaczmarz showed (see [55]) that for the Cesàro methods $T = (C, \alpha)$, the nondecreasing majorants of the Lebesgue functions $L_n(T, x)$ are Weyl multipliers for the summability almost everywhere. Analogous results are true for the methods of Riesz (Sunouchi [130], Leindler [70]), de la Vallée-Poussin (Efimov [29]), and also for the methods $T[v_n]$, $v_1 < v_2 < \cdots$ (Alexits [1], Leindler [68]). The last methods are defined by the matrices $\|\alpha_{nj}\| : \alpha_{nv_n} = 1$, $\alpha_{nj} = 0$ for all other j. Summability by the method $T[v_n]$ is equivalent to the convergence of the sequence S_{v_n} of partial sums.

For general methods of summation, however, these results do not hold (Efimov [28]). Ulyanov pointed out [158] a simple proof of this fact: let $\sum c_k \phi_k$ be an almost everywhere divergent series in L^2 with respect to a bounded ONS. Some sequence $\omega(n) \nearrow \infty$ satisfies the condition $\sum c_n^2 \omega(n) < \infty$. Consider the method $T[v_n]$, where $\{v_n\}$ is a nondecreasing sequence that ranges through all natural numbers. If each of these is taken sufficiently many times, then the condition $L(T, x) = L_{v_n}(x) < \omega(n)$ is satisfied. At the same time, the sequence $\sigma_n = S_{v_n}$ diverges almost everywhere.

The case associated with Theorem 1 is more complicated; that is, in the case where the majorants $\omega(n)$ are bounded. Móricz and Tandori showed [80] that in this case also there exists a counterexample: *there can be constructed an orthonormal system* ϕ *and a regular method of summation* T *such that the condition*

$$L_n(T,x) = O_x(1) \quad (x \in [a,b])$$

is satisfied, and at the same time, some Fourier series from L^2 *is not summable by this method almost everywhere.*

For the proof see below (Theorem 3).

The function $\sup_n L_n(T,x)$ in this example can be made to be summable; however, it is essentially unbounded. Thus, there remains the question of whether it is possible to construct such an example satisfying condition (8).

For general singular integrals, not connected with orthogonal systems, an example of this type was indicated by Zahorski [169]: he constructed a singular kernel $K_n(x,t)$ (that represents an averaging on various intervals) that satisfies the condition $\int_0^1 |K_n(x,t)| dt \le K$ $(x \in [0,1], n=1,2,...)$ (this is the analogue of condition (8)), and for some function $f \in L^x$, the integral $\int_0^1 K_n(x,t) f(t) dt$ diverges as $n \to \infty$ on a set of positive measure.

Summability and convergence of subsequences. The methods $T[v_n]$ referred to above play a useful role in the general theory of the summability of Fourier series from L^2. They represent various concrete summation processes. For example, the method of arithmetic means is *equivalent in the class of* L^2-*series* to the method $T[2^n]$ (Kolmogorov, Kaczmarz, see [55]). This means that if a series (6) with $\sum c_n^2 < \infty$ is summable $(C,1)$ on some set E, then the sequence of partial sums $S_{2^n}(x)$ converges almost everywhere on this set, and vice versa. The result holds also for summation by (C,α), $\alpha > 0$ (Zygmund [171]), and for a wide class of methods that arise from summator functions (Efimov [30]). Similar results hold for the methods of Borel and Euler (Ziza [170]). In this case, $v_n = n^2$, whence in particular it follows that an exact Weyl multiplier for summation by these methods is not better than that for convergence: $\omega(n) = \ln^2 n$.

For arbitrary regular methods it is known that each of them can be contained in some method $T[v_n]$ (Kaczmarz, see [55]). More precisely, for any regular method T there exists a sequence $v_n \uparrow$, such that for each orthogonal series from L^2 that is T-summable on the set E, the subsequence of partial sums S_{v_n} converges almost everywhere on E. However, generally speaking, equivalence no longer holds in this case. Specifically, D. E. Menshov *constructed an orthonormal uniformly bounded system* $\{\phi_n\}$ *and a regular positive method* T *that is not equivalent, in the class of* L^2-*series with respect to this system, to any method* $T[v_n]$. This result forms the subject of the work [76].

Below is a proof of a theorem that synthesizes the theorem of Menshov and the above-mentioned result of Móricz and Tandori. The use of the ω-ordering of the Haar system (see Chap. III) greatly simplifies the construction, in comparison with the works cited, and perhaps makes the essence of the phenomenon more transparent.

Theorem 3. *There exist a complete ONS ϕ and a positive regular method of summation T that have the following properties:*

(i) $L_n(T,x)=O_x(1)$ $(\forall x)$;

(ii) *for some function $f\in L^2$, the Fourier series (6) is not summable by the method T almost everywhere;*

(iii) *the method T is not equivalent, in the class of series (6) from L^2, to any method $T[v_n]$.*

It is clear that (ii) in essential part follows from (iii).

We introduce some notation:

$$m_k = 4^k; \quad \Delta_k = \left[0,\frac{1}{2^{2m_k-1}}\right]; \quad i_k = 2^{2m_k} - 2^{m_k}; \quad I_k = 2^{4m_k} - 2^{m_k}; \quad N_k = \sum_{q=1}^{k} I_q.$$

Let $\{\psi_k^{(i)}\}$ $(1\leq i\leq i_k)$ be the ω-ordered collection of Haar functions $\{\chi_m^{(j)}\}$ $(1\leq j\leq 2^m, m_k\leq m<2m_k)$. In the subspace Φ_k of dimension I_k generated by the functions $\{\chi_m^{(j)}\}$ $(1\leq j\leq 2^m, m_k\leq m<4m_k\equiv m_{k+1})$ we pass to a new orthonormal basis $\{\phi_k^{(i)}\}$ $(1\leq i\leq I_k)$:

$$\phi_k^{(i)} = \frac{1}{\sqrt{2}}(\psi_k^{(i)} + \chi_{4m_k-1}^{(i)}), \qquad \phi_k^{(2i_k+1-i)} = \frac{1}{\sqrt{2}}(\psi_k^{(i)} - \chi_{4m_k-1}^{(i)}) \quad (1\leq i\leq i_k)$$

(notice that the second expression goes to zero outside Δ_k); we take all the other functions $\chi_m^{(j)}\in\Phi_k$ to be $\phi_k^{(i)}$ $(2i_k<i\leq I_k)$. The system $\phi = \left(\bigcup_{i=1}^{16}\chi_i\right)\cup\left(\bigcup_{k=1}^{\infty}\bigcup_{i=1}^{I_k}\phi_k^{(i)}\right)$ is the one we are looking for.

Let there be associated to each n the numbers k and l determined by the conditions $N_{k-1}<n\leq N_k$, $l=n-N_{k-1}$. Following [76], we define a positive regular method of summation $T=\|\alpha_{n,j}\|$:

$$\alpha_{n,n} = \alpha_{n,N_{k-1}+2i_k-l} = \tfrac{1}{2} \quad (l<i_k),$$

$$\alpha_{n,N_{k-1}} = 1 \qquad\qquad (l\geq i_k),$$

with all the other elements of the matrix T equal to zero.

We shall verify condition (i). For $N_{k-1}+i_k\leq n\leq N_k$, we have

$$L_n(T,x) = L_{N_{k-1}}^{\phi}(x) = L_{N_{k-1}}^{\chi}(x) = 1. \tag{9}$$

Further, for $x,t\notin\Delta_k$, $l<i_k$ we have

$$\sum_{i=1}^{l}\phi_k^{(i)}(x)\phi_k^{(i)}(t) + \sum_{i=i_k+1}^{2i_k-l}\phi_k^{(i)}(x)\phi_k^{(i)}(t) = \tfrac{1}{2}\sum_{i=1}^{l}\psi_k^{(i)}(x)\psi_k^{(i)}(t) + \tfrac{1}{2}\sum_{i=l+1}^{i_k}\psi_k^{(i)}(x)\psi_k^{(i)}(t)$$

$$= \tfrac{1}{2}\sum_{m=m_k}^{2m_k-1}\sum_{j=1}^{2^m}\chi_m^{(j)}(x)\chi_m^{(j)}(t).$$

Therefore for $N_{k-1}<n<N_k+i_k$, $x\notin\Delta_k$, we have

$$L_n(T,x) = \tfrac{1}{2}\int_0^1\left|2\left[\sum_{i=1}^{16}\chi_i(x)\chi_i(t) + \sum_{q=1}^{k-1}\sum_{i=1}^{I_q}\phi_q^{(i)}(x)\phi_q^{(i)}(t)\right] + \sum_{i=1}^{i_k}\phi_k^{(i)}(x)\phi_k^{(i)}(t)\right.$$

$$\left. + \sum_{i\in[1,l]\cup[i_k+1,2i_k-l]}\phi_k^{(i)}(x)\phi_k^{(i)}(t)\right|dt \leq 1 + \tfrac{1}{2}\int_{\Delta_k}\left|\sum_{i=1}^{i_k}\phi_k^{(i)}(x)\phi_k^{(i)}(t)\right| \tag{10}$$

$$+ \sum_{i\in[1.l]\cup[i_k+1.2\,i_k-l]} \phi_k^{(i)}(x)\,\phi_k^{(i)}(t)\Bigg|\,dt + \tfrac{1}{2}\int_{C\Delta_k}|...|\,dt \le 2 + \tfrac{1}{2}\int_{\Delta_k}|...|\,dt.$$

Notice that for $x\notin\Delta_k$, for any subset of indices $w\subset[1.2\,i_k]$,

$$\int_{\Delta_k}\Big|\sum_{i\in w}\phi_k^{(i)}(x)\,\phi_k^{(i)}(t)\Big|\,dt \le \sqrt{|\Delta_k|}\Big[\int_{\Delta_k}\Big|\sum_{i\in w}\phi_k^{(i)}(x)\,\phi_k^{(i)}(t)\Big|^2\,dt\Big]^{1.2} \le \sqrt{|\Delta_k|}\Big\{\sum_{i=1}^{2\,i_k}[\phi_k^{(i)}(x)]^2\Big\}^{1.2}$$

$$= \frac{1}{2^{m_k-1.2}}\cdot\frac{1}{\sqrt{2}}\Big\{\sum_{m=m_k}^{2\,m_k-1}\sum_{j=1}^{2^m}[\chi_m^{(j)}(x)]^2\Big\}^{1/2} \le \frac{2^{m_k}}{2^{m_k-1}}.$$

Comparing this with (9) and (10). we see that for any $x\in(0.1]$. starting with some number $n=n(x)$, the inequality $L_n(T,x)<4$ is satisfied.

Now let

$$f = \sum_k \frac{1}{m_k}\sum_{i=1}^{i_k}\|\psi_k^{(i)}\|_1\,\phi_k^{(i)} \equiv \sum_k\sum_i c_k^{(i)}\,\phi_k^{(i)}. \tag{11}$$

Then

$$\sum_k\sum_i|c_k^{(i)}|^2 = \sum_k\frac{1}{m_k^2}\sum_i\|\psi_k^{(i)}\|_1^2 = \sum_k\frac{1}{m_k^2}\sum_{m=m_k}^{2\,m_k-1}\sum_{j=1}^{2^m}\|\chi_m^{(j)}\|_1^2 = \sum\frac{1}{m_k} < \infty.$$

Therefore $f\in L^2$. At the same time, for $N_{k-1}<n<N_{k-1}+i_k$, we have

$$\sigma_n(f) = \tfrac{1}{2}[S_n(f)+S_{N_{k-1}+2\,i_k-l}(f)] = \tfrac{1}{2}\Big[S_{N_{k-1}}(f) + \sum_{i=1}^l c_k^{(i)}\,\phi_k^{(i)} + S_{N_k}(f)\Big].$$

Taking inequality (4) § 2. Chap. III into account. we obtain, for any dyadic-irrational $x\notin\Delta_k$, that

$$\max_{1\le p\le i_k}\Big|\sum_{i=1}^p c_k^{(i)}\,\phi_k^{(i)}(x)\Big| = \max_{1\le p\le i_k}\frac{1}{\sqrt{2}\,m_k}\Big|\sum_{i=1}^p\|\psi_k^{(i)}\|_1\,\psi_k^{(i)}(x)\Big| \ge \tfrac{1}{8}.$$

Thus. the T-means of series (11) diverge almost everywhere.

For the proof of assertion (iii). we fix an arbitrary sequence of indices $v_r\uparrow$. Two cases are possible:

(a) the subsequence $S_{v_r}(f)$ of partial sums of series (11) converges almost everywhere;

(b) the indicated subsequence diverges on a set E, $\mu E>0$.

In case (a). we see that series (11) is summable almost everywhere by the method $T[v_r]$ and not summable by the method T; that is. these methods are not equivalent. In case (b). we divide the sequence $\{v_r\}$ into two subsequences $\{v_r'\}$ and $\{v_r''\}$, where $l(v_r')<i_k$. $l(v_r'')\ge i_k$. Clearly $S_{v_r''}(f)=S_{N_k}(f)\to f$ a.e. Therefore $S_{v_r'}$ diverges almost everywhere on E. Let

$$\tilde f = \sum_k\sum_{i=i_k+1}^{2\,i_k} c_k^{(2\,i_k+1-i)}\,\phi_k^{(i)}.$$

We have

$$S_{v_r'}(f+\tilde f) = S_{v_r'}(f) + S_{N_{k(v_r')}-1}(\tilde f).$$

The second term converges almost everywhere to $\tilde{f}$. Therefore $S_{v_r}(f + \tilde{f})$ diverges almost everywhere on E; that is, the Fourier series of $f + \tilde{f}$ is not summable by the method $T[v_r]$. At the same time, this series is summable by the method T almost everywhere on $[0,1]$. Indeed, for $l(n) < i_k$ we have

$$\sigma_n(f + \tilde{f}) = \tfrac{1}{2}\left[S_{N_{k-1}}(f + \tilde{f}) + S_{N_{k-1}+i_k}(f + \tilde{f}) + \sum_{i=1}^{l} c_k^{(i)} \phi_k^{(i)} + \sum_{i=i_k+1}^{2i_k-l} c_k^{(2i_k+1-i)} \phi_k^{(i)} \right].$$

The first two terms tend to $f + \tilde{f}$ almost everywhere. The sum of the last two terms (for $x \notin \Delta_k$) is equal to

$$\sum_{i=1}^{i_k} c_k^{(i)} \phi_k^{(i)} = S_{N_k}(f) - S_{N_{k-1}}(f) = o(1) \quad \text{a.e.}$$

Thus, the Fourier series of the function $f + \tilde{f}$ is summable almost everywhere by the method T, but not summable be the method $T[v_r]$ almost everywhere on E. Therefore, in case (b) also, these methods are not equivalent. This completes the proof of Theorem 3.

The application of the matrices from § 1, according to the same scheme as in Theorem 1, § 4 (see below) allows us to make the system ϕ uniformly bounded.

Appendix. In conclusion, we mention a generalization of relation (5) to nonorthogonal series, which was shown by Alexits and Sharma [2].

Let $\{\phi_n\}$ be an arbitrary system of functions $\in L^2[a,b]$. It is possible to define Lebesgue functions for this system by means of (1). The following statement is true: *if condition (3) is satisfied, then for any series*

$$\sum c_n \phi_n(x), \qquad \sum c_n^2 < \infty, \tag{12}$$

the partial sums satisfy estimate (5).

In particular, if the Lebesgue functions are bounded, then series (12) converges almost everywhere.

We shall indicate a simple proof of the theorem stated. It is sufficient to be convinced that inequality (4) is preserved in the general case. From the Schur theorem (see § 4, Chap. II) it follows that a finite collection of functions $\{\phi_i\}$ $(1 \le i \le n)$ can be extended on an interval $[b, b+\delta]$ to an orthogonal system equinormed on $[a, b+\delta]$ (the latter means that $\|\phi_i\|_2 = \lambda(n)$ $(1 \le i \le n)$). $\delta > 0$ can be chosen arbitrarily. The Lebesgue functions $L_k(x)$, $x \in [a,b]$ increase after this continuation by the quantity

$$\int_b^{b+\delta} \left| \sum_{i=1}^{k} \phi_i(x) \phi_i(t) \right| dt \le \sqrt{\delta} \left\{ \int_b^{b+\delta} \left| \sum_{i=1}^{k} \phi_i(x) \phi_i(t) \right|^2 dt \right\}^{1/2}.$$

Taking into consideration that the second factor on the right in fact does not depend on δ, we conclude that for sufficiently small δ the system $\{\phi_n\}$ $(1 \le k \le n)$ considered on $[a, b+\delta]$ will satisfy (3) on a set $E_\delta \subset E$ that differs in measure by an arbitrarily small amount from E. Finally observe that because inequality (4) is homogeneous it is preserved for equinormed orthogonal systems.

§ 3. Convergence of Fourier Series of Functions from Various Classes

A system of functions $\phi = \{\phi_n\}$ orthonormal on $[a,b]$ is called a *system of convergence in the class* $A \subset L[a,b]$ if for any function $f \in A$ the Fourier series

$$f \sim \sum c_n(f)\phi_n \tag{1}$$

converges almost everywhere.

The trigonometric system is a system of convergence in $L^p, p > 1$ (Carleson, Hunt) but not in L (Kolmogorov).

Analogous results hold for the Walsh system (Billard [11], Stein [128]). The Haar system is a system of convergence in L.

By means of small perturbations of the trigonometric system, it is not difficult to construct for any p, p', $1 < p' < p < 2$, examples of systems that are systems of convergence in L^p but not in $L^{p'}$ (see below). This is not surprising since the classes L^p, $p < 2$, differ in the rate of decrease of their Fourier coefficients (the Hausdorff-Young and Paley inequalities, see [55]).

The situation is somewhat different for $p \geq 2$. In this case the classes L^p are very similar from the point of view of the properties of their coefficients. This is indicated, for example, by the results about Carleman singularities for complete systems (Chap. III, § 4). This similarity is especially clear for bounded systems: for any sequence $\{b_n\} \in l_2$ there exists $F \in \bigcap_{p < \alpha} L^p$ for which $|c_n(F)| = |b_n|$ (p. 47).

The similarity in the properties of the Fourier series from the classes L^p, $p \geq 2$, is shown also in the questions of convergence (see, for example, § 2, Chap. III). Notice that the theorem of Kolmogorov-Seliverstov-Plessner-Kaczmarz (p. 100) shows that if for any continuous function the Fourier series with respect to a given system converges everywhere, then for any function $f \in L^2$ convergence almost everywhere occurs.

In this connection a question arises: is every system of convergence in C (or in L^p, $2 < p < \infty$) also a system of convergence in L^2? It turns out that in the general situation the answer is negative, even for bounded systems.

Theorem 1 [107]. *Let $p_0 \in [1, \infty)$ be given. Then there exists a complete uniformly bounded ONS Φ having the following properties:*

(i) $\forall f \in L^p$, $p > p_0$, *the Fourier series* (1) *converges to f almost everywhere;*

(ii) $\exists F \in L^{p_0}$ *for which the series* (1) *diverges almost everywhere.*

In the case $p = 1$, the problem is solved by the trigonometric system.

Let $1 < p_0 < 2$. Suppose $\phi_k = \sqrt{\dfrac{2}{\pi}} \sin kx$ $(x \in [0,\pi], k = 1,2,\ldots)$, $\tilde{\psi}_k = \phi_{2^k}$ and $\{\psi_k\}$ is the system consisting of all the other functions of the system ϕ. Assume further that

$$\theta_k = \lambda_k[\psi_k + \varepsilon \tilde{\psi}_k], \qquad \tilde{\theta}_k = \lambda_k[\tilde{\psi}_k - \varepsilon_k \psi_k],$$

where $\varepsilon_k = \ln k \cdot k^{1/2 - 1/p_0}$, $\lambda_k = (1 + \varepsilon_k^2)^{-1/2}$.

For $f \in L^p$, $p > 1$, we define $c_k(f) = (f, \psi_k)$, $\tilde{c}_k(f) = (f, \tilde{\psi}_k)$. Then $\sum \tilde{c}_k^2 < \infty$ (see [171] Chap. V, § 6). The Fourier series of the function f with respect to the systems θ and $\tilde{\theta}$ are written in the following forms:

$$\sum (f, \theta_k) \theta_k = \sum \lambda_k^2 c_k \psi_k + \sum \lambda_k^2 \varepsilon_k \tilde{c}_k \psi_k + \sum \lambda_k^2 \varepsilon_k^2 \tilde{c}_k \tilde{\psi}_k + \sum \lambda_k^2 \varepsilon_k c_k \tilde{\psi}_k;$$

$$\sum (f, \tilde{\theta}_k) \tilde{\theta}_k = \sum \lambda_k^2 \varepsilon_k^2 c_k \psi_k - \sum \lambda_k^2 \varepsilon_k \tilde{c}_k \psi_k + \sum \lambda_k^2 \tilde{c}_k \tilde{\psi}_k - \sum \lambda_k^2 \varepsilon_k c_k \tilde{\psi}_k. \qquad (2)$$

From the theorem of Carleson and Hunt, it is easy to conclude that the first three sums in each of these series converge almost everywhere (it must be observed that the monotone bounded sequences $\{\lambda_k^2\}$, $\{\lambda_k^2 \varepsilon_k^2\}$ are multipliers for convergence almost everywhere, because of the Abel transformation).

In connection with the last sums in (2), observe that their convergence or divergence is dependent on the fulfillment of the condition

$$\sum c_k^2(f) \varepsilon_k^2 < \infty \qquad (3)$$

(this follows from the theory of lacunary trigonometric series; see [171]).

From the Hausdorff-Young inequality we have

$$\sum c_k^2(f) \varepsilon_k^2 \le \left(\sum |c_k|^q \right)^{\frac{2}{q}} \left(\sum \varepsilon_k^{\frac{2q}{q-2}} \right)^{\frac{q-2}{q}} \le K(p) \|f\|_p^2 \left(\sum \varepsilon_k^{\frac{2q}{q-2}} \right)^{\frac{q-2}{q}} \quad \left(q = \frac{p}{p-1} \right).$$

It is immediately verified that for $p > p_0$ the last sum is finite; that is, the series (2) converge almost everywhere for $f \in L^p$.

Consider the function

$$F = \sum k^{-\frac{1}{q_0}} (\ln k)^{-1} \psi_k.$$

It is easy to see that $\sum |c_k(F)|^{p_0} k^{p_0 - 2} < \infty$, whence by the theorem of Hardy and Littlewood ([171] Chap. XII, § 6) we conclude that $F \in L^{p_0}$. At the same time, relation (3) for this function is violated.

It is easy to see now that by uniting the systems θ and $\tilde{\theta}$ (separating the elements $\tilde{\theta}_k$ by sufficiently large intervals), we obtain the required system Φ.

Observation. The construction demonstrated above gives us *an example of a Fourier series from L^p with respect to a complete bounded system for which any subsequence of partial sums diverges almost everywhere.* The possibility of such a phenomenon for $p < 6/5$ was discovered by Marcinkiewicz (see [47]). As we see, it is possible for any $p < 2$, and it can occur while the Fourier series from the classes $L^{p+\varepsilon}$ have good properties.

At the same time, as we know, every Fourier series from L^2 has an almost everywhere convergent subsequence of partial sums S_{v_r}, where $\{v_r\}$ depends only on the system (see [47]). This means that for each system there is a regular method T such that the L^2-series with respect to this system are summable almost everywhere. Orlicz showed [47] that this result does not extend to L^p for $p < 2$: there exists an orthonormal system having the property that for any regular method T there can be found a function $f \in \bigcap_{p < 2} L^p$ (depending on T) whose Fourier series (1) is not summable almost everywhere. The construction given above, for an appropriate ordering of the system $\Phi = \theta \cup \tilde\theta$, gives a stronger example: *there exists a fixed Fourier series* (1) *of a function* $f \in \bigcap\limits_{p<2} L^p$ *that is not summable by any regular method.*

We also mention that the system Φ just constructed forms a basis in L^p, $p_0 < p \leq 2$, but not for $p = p_0$. This follows immediately from the proof above if we keep in mind the theorem of M. Riesz: the trigonometric system is a basis in all L^p, $1 < p < \infty$. In connection with this theorem we mention an unexpected result that was established recently by Fefferman [32]: *the double trigonometric system* $\{e^{i(n,x)}\}$, $n = (n_1, n_2)$, $x = (x_1, x_2)$, *does not form a basis in the space* L^p *for any* $p \neq 2$. Here we have in mind the spherical partial sums

$$S_\rho(f; x) = \sum_{|n| \leq \rho} c_n e^{i(n,x)}.$$

Proof of Theorem 1 (fundamental case $p_0 \geq 2$). We divide the construction into two stages: first we shall define a system ϕ having all the required properties, except boundedness.

We introduce some notation:

$$\Delta_k = [2^{-k}, 1]; \quad \Delta'_k = [0, 2^{-k}]; \quad \gamma_k = 2^{k\left(\frac{1}{2} + \frac{1}{\sqrt{k}} - \frac{1}{p_0}\right)}; \quad \varepsilon_k = 2^{-k}\gamma_k; \quad v_k = 4^k. \tag{4}$$

The system ϕ is constructed from the blocks $\phi^{(k)}$, each of which is equivalent to some collection of Haar functions. We shall describe the construction of the k-th block. Let m_{k-1} be the largest subscript of the Haar functions that take part in the construction of the $(k-1)$-th block. We denote by $\{\tilde\psi_k^{(i)}\}$ $(1 \leq i \leq i_k)$ the finite Haar subsystem consisting of all the functions $\{\chi_m^{(j)}\}$ $(m_k < m \leq m_k + v_k)$ whose support lies in the interval Δ_k. We shall denote this same collection, after the permutation ω (see Chap. III, § 2), by $\{\psi_k^{(i)}\}$. Suppose further that $\alpha_k = 2^{-k}|\Delta_k|$ and denote by $\{\tilde\theta_k^{(i)}\}$ $(1 \leq i \leq i_k)$ the Haar subsystem consisting of all the functions $\{\chi_m^{(j)}\}$ $(m_k + k < m \leq m_k + v_k + k \equiv m_{k+1})$ with support in $\Delta''_k = [0, \alpha_k]$. Let $\{\theta_k^{(i)}\} = \omega\{\tilde\theta_k^{(i)}\}$. It is easy to see that if U_k denotes the unitary operator arising from the linear transformation τ_k of the interval Δ_k onto Δ''_k, $(\tau_k(2^{-k}) = 0)$, that is,

$$U_k f(x) = \left(\frac{|\Delta_k|}{|\Delta''_k|}\right)^{1/2} f(\tau_k^{-1} x), \text{ then we have the equality}$$

$$\theta_k^{(i)} = U_k \psi_k^{(i)} \quad (1 \leq i \leq i_k). \tag{5}$$

Let

$$\phi_k^{(i)} = \lambda_k [\theta_k^{(i)} + \varepsilon_k \psi_k^{(i)}]; \quad \phi_k^{(i+i_k)} = \lambda_k [\varepsilon_k \tilde\theta_k^{(i)} - \tilde\psi_k^{(i)}], \tag{6}$$

where λ_k are normalizing constants $\left(\frac{1}{\sqrt{2}} < \lambda_k < 1\right)$. For

$$\phi_k^{(i)} \ (2 i_k < i \leq 2^{m_{k+1}+1} - 2^{m_k+1} \equiv I_k)$$

we take all the other functions $\{\chi_m^{(j)}\}$ $(m_k < m \leq m_{k+1})$. The block $\phi^{(k)}$ will now consist of the collection $\{\phi_k^{(i)}\}$ $(1 \leq i \leq I_k)$. Letting $\phi = \bigcup \phi^{(k)}$ (with the natural ordering), we obtain an orthonormal system equivalent to the Haar system. Define

$$\delta_k(f;x) = \max_{1 \le l \le i_k} \left| \sum_{i=1}^{l} c_k^{(i)}(f) \phi_k^{(i)}(x) \right|, \quad c_k^{(i)} = (f, \phi_k^{(i)}).$$

We will verify the following statements:

(I) If $f \in L^2$, $f(x) = 0$ $(x \in \Delta_k')$, then $\int_{\Delta_k} \delta_k^2(f;x) dx \le \sum_{i=1}^{i_k} (f, \psi_k^{(i)})^2$ $(k > k(p_0))$. Indeed, for $x \in \Delta_k$ we have

$$\delta_k^2(f;x) \le \left(\sum_{i=1}^{i_k} |c_k^{(i)} \phi_k^{(i)}(x)| \right)^2 = \lambda_k^4 \varepsilon_k^4 \left(\sum \left| \int_{\Delta_k} f \psi_k^{(i)} dt \right| |\psi_k^{(i)}(x)| \right)^2$$

$$\le \varepsilon_k^2 v_k \sum |(f, \psi_k^{(i)})|^2 |\psi_k^{(i)}(x)|^2$$

(here we have used the fact that for each x no more than v_k terms can be different from zero). Hence after integrating and using (4) we obtain the required inequality.

(II) If $f \in L^p$, $f(x) = 0$ $(x \in \Delta_k)$, then

$$\int_{\Delta_k} \delta_k(f;x) dx \le \|f\|_p 2^{k\left(\frac{1}{p} + \frac{1}{\sqrt{k}} - \frac{1}{p_0}\right)}.$$

Indeed, taking into account the fact that $\|\chi_m^{(j)}\|_1 = 2^{-m/2}$, and using the Hölder inequality, we obtain

$$\int_{\Delta_k} \delta_k dx \le \lambda_k^2 \varepsilon_k \int_{\Delta_k} \sum |(f, \theta_k^{(i)})| |\psi_k^{(i)}(x)| dx \le \varepsilon_k \left(\sum (f, \theta_k^{(i)})^2 \right)^{1/2} \left(\sum \|\psi_k^{(i)}\|_1^2 \right)^{1/2}$$

$$\le \varepsilon_k \left(\int_{\Delta_k'} f^2 dx \right)^{1/2} \sqrt{v_k} \le \gamma_k \|f\|_p |\Delta_k'|^{1/2 - 1/p}.$$

(III) Suppose $F_k = \frac{1}{\gamma_k \sqrt{v_k}} \sum_{i=1}^{i_k} \|\psi_k^{(i)}\|_1 \theta_k^{(i)}$. Then,

$$\|F_k\|_{p_0} \le C(p_0) 2^{-\sqrt{k}}, \quad \delta_k(F_k;x) \ge \tfrac{1}{8} \quad \text{(for almost every } x \in \Delta_k). \tag{7}$$

Indeed, taking (5) into consideration, we have

$$\|F_k\|_p = \frac{1}{\gamma_k \sqrt{v_k}} \|U_k(\sum \|\psi_k^{(i)}\|_1 \psi_k^{(i)})\|_{L^p(\Delta_k')}$$

$$= \frac{1}{\gamma_k \sqrt{v_k}} \|\sum \|\psi_k^{(i)}\|_1 \psi_k^{(i)}\|_{L^p(\Delta_k)} \cdot \left(\frac{|\Delta_k|}{|\Delta_k''|} \right)^{\frac{1}{2} - \frac{1}{p}} \le 2^{k\left(\frac{1}{p_0} - \frac{1}{p} - \frac{1}{\sqrt{k}}\right)} \frac{1}{\sqrt{v_k}} \left\| \sum_{m=m_k+1}^{m_k+v_k} \mathfrak{r}_{m+1} \right\|_p.$$

Hence, because of the Khinchin inequality for the Rademacher system, we obtain the first of the inequalities in (7). Finally, according to inequality (4) (§ 3, Chap. II) we obtain, for dyadic-irrational $x \in \Delta_k$,

$$\delta_k(F_k;x) = \frac{1}{\gamma_k \sqrt{v_k}} \max_{1 \le l \le i_k} \left| \sum_{i=1}^{l} \lambda_k^2 \|\psi_k^{(i)}\|_1 \varepsilon_k \psi_k^{(i)}(x) \right| \ge \frac{\varepsilon_k}{2\gamma_k \sqrt{v_k}} \max_{1 \le l \le i_k} \left| \sum_{i=1}^{l} \|\psi_k^{(i)}\|_1 \psi_k^{(i)}(x) \right|$$

$$\ge \frac{1}{8 v_k} \sum_{i=1}^{i_k} \|\psi_k^{(i)}\|_1 |\psi_k^{(i)}(x)| = \frac{1}{8 v_k} \sum_{m=m_k+1}^{m_k+v_k} |\mathfrak{r}_{m+1}(x)| = \tfrac{1}{8}.$$

By now it is easy to see that the system ϕ satisfies conditions (i) and (ii) of the theorem. In fact, inequality (7) shows that the function $F=\sum F_k \in L^{p_0}$, while its Fourier series

$$\sum d_i \phi_i = \sum_k \left(\sum_{i=1}^{i_k} c_k^{(i)}(F) \phi_k^{(i)} + \sum_{i=i_k+1}^{2i_k} c_k^{(i)} \phi_k^{(i)} \right)$$

diverges almost everywhere. Now let $f \in L^p$ for some $p > p_0$. We shall show that its Fourier series converges almost everywhere. This series is broken into three subseries:

$$\sum_k \sum_{i=1}^{i_k} c_k^{(i)} \phi_k^{(i)}, \quad \sum_k \sum_{i=i_k+1}^{2i_k} c_k^{(i)} \phi_k^{(i)}, \quad \sum_k \sum_{i=2i_k+1}^{l_k} c_k^{(i)} \phi_k^{(i)},$$

and the convergence of the last two is ensured by the condition that $f \in L^2$, because of the properties of the Haar system. This is also true for the series $\sum g_k$, where $g_k = \sum_{i=1}^{i_k} c_k^{(i)} \phi_k^{(i)}$. We have only to verify that $\delta_k(f; x) \to 0$. Let $f = f_k + f_k'$, $f_k(x)=0$ $(x \in \Delta_k')$, $f_k'(x)=0$ $(x \in \Delta_k)$. If we fix $k_0 > k(p_0)$, then, on the basis of (I) and (II), we have

$$\sum_{k>k_0} \int_{\Delta_{k_0}} \delta_k^2(f_k) dx \le \sum_{k>k_0} \int_{\Delta_k} \delta_k^2(f_k) dx \le \sum_{k>k_0} \sum_{i=1}^{i_k} (f_k, \psi_k^{(i)})^2 \le \|f\|_2^2,$$

$$\sum_{k>k_0} \int_{\Delta_{k_0}} \delta_k(f_k') dx \le \sum_{k>k_0} \|f_k'\|_p \cdot 2^{-k\left(\frac{1}{p_0}-\frac{1}{p}-\frac{1}{\sqrt{k}}\right)} < \infty.$$

Hence $\delta_k(f; x) \le \delta_k(f_k; x) + \delta_k(f_k'; x) = o_x(1)$ almost everywhere on Δ_k, and this means also on $[0,1]$.

It is necessary for us now to convert the system ϕ into a uniformly bounded one without losing the other properties. This achieved by the use of the transformations A_k.

First of all we observe that in the previous construction we can at each step, after the construction of the block $\phi^{(k)}$, omit one group of Haar functions. As a result, the system ϕ (no longer complete) turns out to be equivalent to a Haar subsystem $\{\chi_m^{(j)}\}$ $(1 \le j \le 2^m)$, where the index m runs through an increasing sequence of natural numbers, omitting an infinite number of values $\{\mu_k\}$. We shall consider the system ϕ to be defined just this way.

We can further define an auxiliary ONS Ψ that is equivalent to the Haar subsystem $\{\chi_{\mu_k}^{(j)}\}$ complementary to ϕ and has the following properties:

(a) Ψ is a system of convergence in L^2.

(b) It contains infinitely many Rademacher functions: $\Psi \supset \{r_{\mu_k}\}$. To prove this, it is sufficient to consider the system defined on p. 22, and to select from it a subsystem $\{\psi_n, 2^{\mu_k} < n \le 2^{\mu_k+1}, k=1,2,...\}$. We combine the systems ϕ and Ψ in blocks $G^{(k)}$ that satisfy the conditions

$$G^{(k)} = \{g_k^{(j)}, 1 \le j \le 2^k\}; \quad \|g_k^{(1)}\|_\alpha \le \sqrt{2^k}; \quad \bigcup_k \bigcup_{j=2}^{2^k} g_k^{(j)} \subset \{r_n\}. \tag{8}$$

This can be done without changing the order of the elements of each system. As a result, the system $g=\bigcup G^{(k)}$ so obtained will satisfy conditions (i) and (ii) of the theorem. Define, finally,

$$\Phi_k^{(i)} = \sum_{j=1}^{2^k} a_{ij}^{(k)} g_k^{(j)} \qquad (1 \leq i \leq 2^k; k=1,2,\ldots). \tag{9}$$

Clearly the system $\Phi = \bigcup \{\Phi_k^{(i)}\}$ is orthonormal and equivalent to the Haar system (in particular, it is closed in C). In addition, from condition (I) and property ii 1, and taking (8) into account, we have

$$\|\Phi_k^{(i)}\|_r \leq \left\|\frac{1}{\sqrt{2^k}} g_k^{(1)}\right\|_\infty + \sum_j |a_{ij}^{(k)}| < C+1.$$

For the completion of the proof of the theorem it is sufficient to verify that for any $f \in L^2$ the Fourier series with respect to the systems Φ and g are equi-convergent almost everywhere. For this in turn it is sufficient to establish the inequality

$$\int_0^1 \max_{1 \leq l \leq 2^k} \left| \sum_{i=1}^l d_i \Phi_k^{(i)}(x) \right|^2 dx \leq K \sum_{i=1}^{2^k} d_i^2 \qquad (\forall \{d_i\}) \tag{10}$$

(where K is an absolute constant).

We have

$$\max_{1 \leq l \leq 2^k} \left| \sum_{i=1}^l d_i \Phi_k^{(i)}(x) \right|^2 \leq \left(\max_l \left| \sum_{i=1}^l d_i \left[a_{i1}^{(k)} \mathbf{r}_1 + \sum_{i=2}^{2^k} a_{ij}^{(k)} g_k^{(j)}(x) \right] \right| \right)$$

$$+ \sum_{i=1}^{2^k} |d_i a_{i1}^{(k)}| |g_k^{(1)}(x) - \mathbf{r}_1(x)|^2 .$$

Denote the quantity in the square brackets by p_i. We observe that $\{p_i\}$ is a collection of orthonormal Rademacher polynomials.

For the proof of inequality (10) it is sufficient to establish exactly such an inequality for the system $\{p_i\}$. This is done (see [98]) with a method used by Erdös for the investigation of lacunary trigonometric series (see [10] p. 260 v. 2). We mention that S. B. Stechkin first showed the applicability of the Erdös method to series with respect to general p-lacunary systems ($p>2$) (the last term means that on the set of all linear combinations the L^2 and L^p norms are equivalent).

Inequality (10) completes the proof of Theorem 1.

For some additional references dealing with the case $p=\infty$ see [162], [64'].

We turn our attention to one of the important steps of this proof—the construction of the system ϕ. The selection of the small multipliers ε_k in (6) is carried out in such a way that the system $\{\varepsilon_k \psi_k^{(i)}\}$ will still be a system of divergence in the class of coefficients l_2, while at the same time $\{\varepsilon_k^2 \psi_k^{(i)}\}$ will be a system of convergence. Thanks to this, the convergence almost everywhere of the Fourier series of any function $f \in L^2$ with respect to the system ϕ depends exclusively on the behavior of this function in an arbitrarily small neighborhood of zero. This idea turns out to be useful in other problems, too (see below, Theorem 2).

Remarks about rearrangement of classical systems. We mention that a system with properties (i) and (ii) cannot be obtained by means of only a rearrangement of the Haar system. The following proposition is true [107]: *an arbitrary fixed rearrangement of the Haar system is a system of convergence in all $L^p, 2 \le p < \infty$, or in none of them.* (*This is true even for $1 \le p \le \infty$* — [63']). This result can be obtained in the following way. Let the Fourier series

$$f \sim \sum c_n \chi_n, \qquad f \in L, \tag{11}$$

after a fixed rearrangement of the terms, α, diverge on a set E, $\mu E > 0$. We can assume that unbounded divergence takes place (this is achieved by multiplying the coefficients of the series by a nondecreasing sequence that tends to ∞ slowly and is constant on sufficiently large intervals).

Fix a number $A > 0$ and define

$$v(x) = \min \{v, \varDelta[\chi_v] \ni x; \exists y \in \varDelta[\chi_v], S_v(y) > A\}$$

(here $\varDelta[\chi_v]$ is the support of the function χ_v, and S_v is the v-th partial sum of series (11)); if the set of numbers that satisfy the conditions given above is empty, then we define $v(x) = \infty$. Consider the series

$$\sum_{n=1}^{\infty} \tilde{c}_n \chi_n(x) \equiv \sum_{1 \le n < v(x)} c_n \chi_n(x). \tag{12}$$

Essentially this is the "stop-time" transformation of martingale theory. In the case given, this means a "purging" of the series—some of the coefficients are replaced by zeros. Clearly the partial sums of the series obtained are uniformly bounded. They converge almost everywhere to some function $f_A \in L^\infty$, and series (12) is its Fourier series.

Let $U_n = \{x; v(x) = n\}$, $\bigcup_{1 \le n < \infty} U_n = U$. Clearly if the set U_n is nonempty then for at least one of the halves $\varDelta'_n$ of the interval $\varDelta[\chi_n]$ we have $|S_n(y)| > A \ (\forall y \in \varDelta'_n)$. Taking into account the relation $S_n(y) = \frac{1}{|\varDelta'_n|} \int_{\varDelta'_n} f dt$, we obtain $\mu U_n \le \frac{2}{A} \int_{\varDelta'_n} |f| dt$. Hence $\mu U \le \frac{2\|f\|_1}{A}$.

Outside the set U, the series (11) and (12) are identical. Therefore, for $A > \frac{2\|f\|_1}{\mu E}$, the Fourier-Haar series of the function f_A, after a rearrangement α, diverges unboundedly on E', $\mu E' > 0$. Using Lemma 2, § 2 Chap. III, we obtain a continuous function with the same property.

We do not know whether an analogous result is true for the trigonometric system in the interval $p \in [2, \infty)$.

This is definitely not the case for $p \in (1, 2]$: *there exists a rearrangement of the trigonometric system which is a system of convergence in L^2, but not in L^p for $p < 2$.*

Indeed, according to the theorem of Orlicz (see Chap. II, § 3, Remark 2), it is possible for any $f \in L^p \backslash L^2$ to divide the trigonometric system τ into two subsystems and reunite them without changing the order of the elements in each of these subsystems in such a way that the Fourier series of f with respect to the resulting system ϕ diverges even in measure. It is clear that such a rearrangement (weak, in the terminology of [154]) maintains the convergence almost everywhere of series from L^2.

We mention further the following proposition.

If a rearranged trigonometric system ϕ is a system of convergence in $L^{p_0}, p_0 < 2$, then it forms a basis in L^p for all $p \in (p_0, 2]$.

Actually, the theorem of Stein (Chap. III, § 3) in this case gives the result that the operator

$$G f = \hat{f} = \sup_n |S_n(f; x)|, \qquad S_n(f) = \sum_1^n c_k(f) \phi_k,$$

is of weak type (p_0, p_0) and $(2,2)$. According to the Marcinkiewicz interpolation theorem this means that the operator G is bounded in L^p for $p \in (p_0, 2)$, and it means that the operators S_n are uniformly bounded for each of these values p. The same argument in an analogous situation was first used by E. M. Nikishin.

The possibility of applying the theorem of Stein here results from a special property of the system ϕ: the operator G commutes with translations. In the general case it is easy to construct a complete ONS that is a system of convergence in L and does not form a basis in any $L^p, p \neq 2$. This is done by a perturbation of the Haar system for which the supports of the perturbed functions contract toward some point. With the help of the matrices B_k according to the scheme of Theorem 6, § 2 Chap. I, *it is possible to convert the system obtained into a uniformly bounded complete system of convergence in $\bigcup_{p > 1} L^p$ that is not a basis in L^p for $p \neq 2$.*

It is interesting that in such examples the system fails to be a basis only because of sets $E \subset [0,1]$ of arbitrarily small measure, as follows from the results of Nikishin; see Chap. III, § 3.

We point out that the connection mentioned above between convergence almost everywhere and in mean is not reversible. The following result is true:

There exists a rearrangement of the trigonometric system which is a basis for all $L^p, 1 < p < \infty$, and which is not a system of convergence even in the class C.

For the proof we divide the trigonometric system τ into blocks $T_k = \{\tau_i; 2^k < i \leq 2^{k+1}\}$. An arbitrary rearrangement of the blocks induces some rearrangement of elements of the system τ. By the theorem of Paley ([171] Chap. XV) for every rearrangement of this type the system remains a basis in $L^p, 1 < p < \infty$. At the same time, as was shown in § 2, Chap. III, there exists a rearrangement of the type indicated for which the Fourier series of some continuous function diverges almost everywhere.

System with Banach type singularities. In conclusion we touch upon one more aspect of the theory of divergent Fourier series.

Banach [9], relying on a construction of Menshov for a system of divergence, showed that for any function $f \in L^2, f \neq 0$, there exists an orthonormal system ϕ for which the Fourier series (1) diverges almost everywhere. As Tsereteli [151] observed not long ago, it is possible to obtain this directly from the theorem of Menshov. Let $\{\psi_n\}$ be Menshov's system and let $F \in L^2$ be a function with a divergent Fourier series. We consider a unitary operator U that makes a rotation in the plane generated by the vectors F and $f (UF = f)$ and that is equal to the identity in its orthogonal complement. A simple calculation shows that the system $\phi_k = U\psi_k$ possesses the required property.

An even more general result [9] is due to Banach: if $\mathscr{K}$ is a compact set (or is the union of a countable number of compact sets) in L^2, then there exists an ONS ϕ with respect to which every function $f \in \mathscr{K}, f \neq 0$, has an almost everywhere divergent Fourier series. For example, this is true for sets of smooth functions or Hölder functions.

A similar result holds also for some more extensive sets. Krantsberg [64'], using the methods of the works [103,107], proved the following theorem.

Theorem 2. *There exists a complete bounded ONS ϕ with respect to which every continuous function $f \neq 0$ has an almost everywhere divergent Fourier series.*

The key point is the following.

Lemma. *Let there be given two sets of positive measure $E, e \subset [a,b]$, $E \cap e = \emptyset$, and a number $\alpha > 0$. Then there exists a collection of functions $\{\phi_i\}$ $(1 \leq i \leq 2n, n = n(\alpha), \phi_i(x) = 0$ for $x \notin E \cup e)$, ON on $[a,b]$, having the property that any function f, $\|f\|_\infty \leq 1$, $f(x) > \alpha (x \in e)$, satisfies the inequality*

$$\delta(f; x) \equiv \max_{1 \leq k \leq n} \left| \sum_{i=1}^{k} c_i \phi_i(x) \right| > \frac{1}{\alpha} - \delta_1(f; x), \quad x \in E \qquad (13)$$

where

$$\int_E \delta_1^2 \, dx \leq \sum_{i=1}^{2n} c_i^2; \quad c_i = (f, \phi_i). \qquad (14)$$

By means of a substitution of variable it is easy to reduce the general case to the following: $E = [0,1]$, $e = (-2\beta, 0)$, $\beta > 0$, $[a,b] \supset [-2\beta, 1]$. Fix m_0 and let

$$v > \frac{2^{16}}{\alpha^8 \beta^2}, \quad \varepsilon = v^{-1/4}.$$

Denote by $\{\psi_i\}$ $(1 \leq i \leq n)$ the functions $\{\chi_m^{(j)}\}$ $(1 \leq j \leq 2^m, m_0 < m \leq m_0 + v)$ after the rearrangement ω. Set

$$\theta_i(x) = \frac{1}{\sqrt{v\beta}} \|\psi_i\|_1 \quad (x \in e' \equiv (-2\beta, -\beta)), \quad 1 \leq i \leq n.$$

Taking an arbitrary interval $e'' \subset (-\beta, 0)$, we can continue the system θ to one orthonormal on $e' \cup e''$. For this it is sufficient (as in § 2, Chap. I) to take the orthonormal system $\{\rho_j\}$ $(1 \leq j \leq n-1)$ concentrated on e'' and to apply to the collection $\{\rho_j\}$ $(1 \leq j \leq n)$, where

$$\rho_n(x) = \beta^{-\frac{1}{2}}, \ x \in e',$$
$$\rho_n(x) = 0, \ x \in e'',$$

an orthogonal matrix whose last column is equal to $a_{in} = \frac{1}{\sqrt{v}} \|\psi_i\|_1$. If e'' is chosen sufficiently small, then we will have the inequality

$$\int_{e''} |\theta_i| \, dx < \frac{1}{2^{m_0 + v} \alpha}. \qquad (15)$$

We extend all the functions ψ, θ to the whole interval $[a,b]$, setting them equal to zero. Now define, by analogy with (6),

$$\phi_i = \lambda[\theta_i + \varepsilon\psi_i], \qquad \phi_{i_0+i} = \lambda[\varepsilon\theta_i - \psi_i] \qquad (1 \le i \le n), \qquad \lambda = (1+\varepsilon^2)^{-1/2}.$$

Let the function f satisfy the conditions of the lemma. Define $\dfrac{1}{\beta}\int_{e'} f\, dx = \mu \ge \alpha$. Suppose $f = f' + f'' + g$ where $f' = f\chi(e')$, $f'' = f\chi(e'')$. Estimating $\delta_1(f;x) \equiv \delta(g;x)$ as in assertion (I) of the proof of Theorem 1, we obtain the following inequality:

$$\int_E \delta_1^2\, dx \le \lambda^4 \varepsilon^4 \nu \sum_{i=1}^{i_0} (f,\psi_i)^2 \le \sum_{i=1}^n c_i^2(f).$$

For f'', because of (15), we have $|c_i(f'')| \le \dfrac{1}{2^{m_0+\nu}\alpha}$; that is, $\delta(f'';x) \le \dfrac{1}{\alpha}\ (x \in E)$. Finally, for f' we have

$$c_i(f') = \int_{e'} f\phi_i\, dx = \lambda\frac{1}{\sqrt{\nu\beta}}\|\psi_i\|_1 \int_{e'} f\, dx = \lambda\sqrt{\frac{\beta}{\nu}}\|\psi_i\|_1\mu.$$

Therefore just as in assertion (III) of Theorem 1 we obtain $(x \in E)$

$$\delta(f';x) = \lambda^2 \varepsilon\sqrt{\frac{\beta}{\nu}}\mu \cdot \max_{1 \le k \le n}\left|\sum_{i=1}^k \|\psi_i\|_1\psi_i(x)\right| > \tfrac{1}{8}\alpha\sqrt{\beta\nu}\,\varepsilon > \frac{2}{\alpha}.$$

On comparing these estimates we see that the lemma has been proved.

Remark. Clearly m_0 could have been chosen so that all the functions ϕ_i on the set E would be "almost" orthogonal to any pre-assigned finite collection of functions. Therefore it is possible by means of small perturbations to achieve orthogonality, without losing the other properties.

To complete the proof of the theorem we construct a sequence of pairs of nowhere dense sets $E_k, e_k \subset [a,b]$ satisfying the following conditions:

(a) $E_k \cap e_k = \emptyset$, $\quad e_k \cap \left(\bigcup_{s<k}(E_s \cup e_s)\right) = \emptyset \quad (k=1,2,\ldots)$,

(b) $\mu E_k \to b - a$,

(c) $e_k \subset \left(x_k - \dfrac{1}{k}, x_k + \dfrac{1}{k}\right); \quad \mu e_k > 0$,

where $\{x_k\}$ is a sequence of numbers everywhere dense in $[a,b]$.

Putting $\alpha_k = \dfrac{1}{k}$ we apply the lemma to the sets E_k, e_k, for each k, and we define the systems $\{\phi_k^{(i)}\}$ $(1 \le i \le 2n_k)$. Because of the Remark and consideration of property (a), we can make the system $\phi = \bigcup\{\phi_k^{(i)}\}$ orthogonal. Now let $f \in C$, $\|f\|_\infty < 1$, $f(x_0) > 0$. Then in some neighborhood of the point x_0 the condition $f(x) > \alpha \equiv \alpha(f)$ is fulfilled. Therefore because of (c), the conditions of the lemma will be satisfied for an infinite number of indices k; that is, inequalities (13) and

(14) are satisfied. The last condition gives $\delta_1^{(k)}(x) = o(1)$. Therefore condition (13) together with (b) means that Fourier series with respect to the system ϕ will diverge almost everywhere.

Completion of the system ϕ and its transformation into a uniformly bounded system is accomplished by the scheme of Theorem 1. Technical difficulties increase, however, since the functions $\phi_k^{(i)}$ here are not step functions.

It is possible to formulate the following problem, which encompasses the questions examined in this section and which is, apparently, difficult in so general a form.

Given two sets $A, B \subset L^2$. Under what conditions does there exist an orthonormal system (complete, bounded) with respect to which every function $f \in A$ has an almost everywhere convergent Fourier series, while the Fourier series of any function $f \in B$ diverges almost everywhere?

Orthogonal bases in L^p. Let $\phi = \{\phi_k\}$ be an orthonormal system forming a basis in L^p for some $p \in [1, \infty)$. This is equivalent to the fulfillment of the conditions:

(i) ϕ is closed in L^p;

(ii) there exists a dual system of functionals $\psi_k \in L^q$, $q = \dfrac{p}{p-1}$, $(\phi_l, \psi_k) = \delta_{kl}$, and the projections $P_n = \sum_1^n (f, \psi_k) \phi_k$ have uniformly bounded norms.

Further let the following condition be satisfied:

(iii) $\phi_k \in L^q$ $(\forall k)$.

Because of the uniqueness of the dual system, which follows from (i), we have $\psi_k = \phi_k (\forall k)$; that is, the expansion with respect to the basis has the form $f = \sum (f, \phi_k) \phi_k$. Hence it follows that the system ϕ is complete in the space L^p. Consequently, the system ϕ is closed in L^r for $r \in [p, q]$, $r \neq \infty$. In particular, the following condition is satisfied:

(iii') the system ϕ is complete in L^2.

It is easy to see also that condition (iii') implies (iii), because of the uniqueness of the system dual to ϕ in L^2. Thus, *in the situation being examined the conditions* (iii) *and* (iii') *are equivalent* (P. L. Ulyanov [159]).

Further, the projection $P_n = \sum_1^n (f, \phi_k) \phi_k$ acting on L^q is dual to the same projection in L^p, so their norms coincide. The Riesz interpolation theorem gives the uniform boundedness of these projections in L^r, $r \in [p, q]$.

It follows from what has been said that *for complete systems orthonormal in* L^2 *the set* Δ_ϕ *of values of* $p \in (1, \infty)$ *for which the system forms a basis in* L^p *is always an interval (closed or open) of the form* (p_0, q_0).

The number p_0 *here can be anything.* This follows, for example, from the construction on p. 108 (see the Observation p. 109). As in the case of convergence almost everywhere, we do not know whether arbitrary p_0 can be obtained by rearranging the trigonometric system.

Note that condition (iii') is essential. A. N. Slepchenko showed by an example [123] that it might not be fulfilled (for $p < 2$): *there exists an ONS incomplete in* L^2 *that forms a basis in* L^p *for a given* $p < 2$. The construction of this system is accomplished by induction. Let the orthonormal vectors ϕ_k $(1 \le k \le n)$ be defined, and suppose they satisfy the conditions:

(a) $\|\phi_k - \chi_k\|_p < 2^{-k}$;

(b) if $\displaystyle\sum_{k=1}^{n} \lambda_k \phi_k \in L^q$, then $\lambda_k = 0$ $(\forall k)$.

From the Hahn-Banach theorem it is easy to see that condition (b) guarantees the denseness of the linear subspace $E_n = \{f \in L^2, (f, \phi_k) = 0, 1 \le k \le n\}$ in L^p. Therefore we can take $\phi_{n+1} \in E_n$, $\|\phi_{n+1} - \chi_{n+1}\|_p < 2^{-(n+1)}$. It is easy to see that this choice can be made so that condition (b) still holds. According to the theorem on stability (p. 37) the system $\{\phi_k\}$ forms a basis in L^p together with the Haar system. At the same time condition (iii), and consequently also (iii'), are violated.

A more general example is when the set Δ_ϕ for an incomplete system ϕ coincides with a given interval $\subset [1, 2)$, see [113'].

§ 4. Sums of Fourier Series

We saw in the investigation of Fourier series from the spaces L^p, $p < 2$, that effects arise that contrast sharply with the properties of Fourier series from L^2. This is indicated, in particular, by the results of § 2 (Theorem 2 and following) and § 3 (Observation). These results deal with divergence of Fourier series.

In this section we point out new peculiarities of convergent L^p-series, with $p < 2$.

Suppose that the Fourier series

$$\sum c_n(f) \phi_n(x), \qquad c_n = (f, \phi_n), \tag{1}$$

converges almost everywhere. Is it possible to assert that

$$\sum c_n \phi_n(x) = f(x) \tag{2}$$

almost everywhere?

Of course the completeness of the system is essential here. Banach noticed that the Fourier series of a continuous function can have a sum that is a discontinuous function (see [55]), however the nature of this phenomenon is connected with the fact that the system he constructed was not complete.

With the condition of completeness, the answer to the question given is affirmative for L^2-series, since in this case equality (2) is satisfied in mean. The result extends to bases in L^p, and also to systems possessing a regular method of summation that sums each Fourier series to its original function, f (for example, the trigonometric system with the method of arithmetic means).

It turns out, however, that in the general case a counterexample is possible: *there exists a complete (bounded) system with respect to which the Fourier series (1) of some function f converges almost everywhere to $g \neq f$.*

Completeness here can be taken in the strongest sense—as closure in the space C.

What is more, it turns out that *a fixed Fourier series, with an appropriate ordering, can represent any measurable function.*

A function series $\sum f_k$ is called universal with respect to rearrangements (more briefly, π-universal) if for any measurable function g (possibly taking infinite values) there exists a rearrangement $\pi = \{n_k\}$ for which $\sum f_{n_k} = g$ almost everywhere. Examples of such series are well known.

The following is true.

Theorem [102]. *There exist a bounded ONS closed in C and a function $f \in \bigcap\limits_{p<2} L^p$ whose Fourier series (1) is π-universal.*

Lemma 1. *Let the numbers $\alpha_k = o(1), \sum |\alpha_k| = \infty$ be given. Then for any increasing sequence of natural numbers $\{m_k\}$ the series*

$$\sum_{k=1}^{\infty} \sum_{j=1}^{2^{m_k}} \sum_{r=0}^{1} (-1)^r \alpha_k \chi(\Delta_{m_k}^{(j)}) \tag{3}$$

is π-universal.

(Here $\Delta_m^{(j)} = \left[\dfrac{j-1}{2^m}, \dfrac{j}{2^m} \right]$ and $\chi(E)$ is the characteristic function). This proposition is a modification of Riemann's theorem on conditionally convergent series.

For the proof it is sufficient to take, for a fixed g, a sequence of step functions g_n converging to g almost everywhere (the g_n are constant on each of the intervals $\Delta_n^{(j)}$). It is not difficult to see that the terms of series (3) can be rearranged so that the partial sums σ_i of the series obtained satisfy the condition $\|\sigma_{i_n} - g_n\|_\infty = o(1)$, where $\{i_n\}$ is some increasing sequence of indices, and for $(i_{n-1} < i \leq i_n)$ the values of $\sigma_i(x)$ go beyond the limits of the interval $[\sigma_{i_{n-1}}(x), \sigma_{i_n}(x)]$ no farther than $\varepsilon_n = o(1)$.

Lemma 2. *There exists a closed system $\{\psi_n\}$, orthonormal on $[0,1]$, that consists of bounded functions and that divides into three subsystems $\{\psi_s^{(j)}\}$ $(0 \leq j \leq 2; s = 1, 2, \ldots)$ with the following properties:*

(a) $|\psi_s^{(0)}(x)| \leq 1$ $(\forall x)$;

(b) $\psi_s^{(1)}(x) = 0$ $(x \notin \Delta_n^{(1)})$, $\sum\limits_s \|\psi_s^{(1)}\|_p < \infty$ $(\forall p < 2)$.

(c) *for some sequence $\gamma_s = o(1)$ the series $\sum\limits_s \sum\limits_{r=0}^{1} (-1)^r \gamma_s \psi_s^{(2)}$ is π-universal.*

The system ψ consists of blocks $\phi^{(k)}$, each of which is a collection of functions $\{\phi_k^{(i)}\}$ $(1 \leq i \leq i_k)$ that is equivalent to the Haar subsystem $\{\chi_m^{(j)}\}$ $(1 \leq j \leq 2^m, m_k \leq m < m_{k+1} = 2m_k + 2)$. We define the k-th block in the following manner.

Let $\rho_k = [(2^{m_k} - 1)^{-1} 2^{m_k}]^{1/2}$;

$$\phi_k^{(i)}(x) = \begin{cases} -\rho_k(2^{m_k} - 1), & (i-1)2^{-m_k} < x < (i-1)2^{-m_k} + 2^{-2m_k} \\ \rho_k & , (i-1)2^{-m_k} + 2^{-2m_k} < x < i 2^{-m_k} \\ 0 & , x \notin \Delta_{m_k}^{(i)} \end{cases}$$

$$(1 \leq i \leq 2^{m_k}).$$

Observe that

$$\phi_k^{(i)} = \beta_k \chi(\Delta_{m_k}^{(i)}), \quad x \notin [(i-1)2^{-m_k}, (i-1)2^{-m_k} + 2^{-2m_k}]. \tag{4}$$

It is easy to see that $\{\phi_k^{(i)}\}$ are orthonormal polynomials in the system $\{\chi_m^{(j)}\}$ $(m_k \leq m < 2m_k)$. Let

$$\phi_k^{(i_k-1)} \equiv \psi_k^{(0)} = \mathfrak{r}_{2m_k}; \quad \phi_k^{(i_k)} \equiv \psi_k^{(1)} = \chi_{2m_k+1}^{(1)}. \tag{5}$$

From the subspace of polynomials $\sum\limits_{m=m_k}^{2m_k+2} \sum\limits_j d_j \chi_m^{(j)}$ that are orthogonal to the functions $\phi_k^{(i)}$ already constructed, we choose an orthonormal basis and consider its elements as $\{\phi_k^{(i)}\}$ for $2^{m_k} < i < i_k - 1$. Now letting $\psi = \bigcup \phi^{(k)}$, we obtain a system with the required properties. Indeed, (5) ensures the fulfillment of conditions (a) and (b). Notice further that the subsystem $\{\psi_n^{(2)}\}$ consists of the functions $\bigcup\limits_k \{\phi_k^{(i)}; 1 \leq i < i_k - 1\}$. To prove property (c) it is sufficient to convince oneself that the series $\sum\limits_k \sum\limits_i \sum\limits_{r=0}^{2^{m_k}} (-1)^r \dfrac{1}{\sqrt{k}} \phi_k^{(i)}$ is π-universal. Because of (4), the terms of this series differ from the corresponding terms of series (3) (where $\alpha_k = \dfrac{\beta_k}{\sqrt{k}}$) on sets $E_k^{(i)}, \mu E_k^{(i)} < 2^{-2m_k}$. Finally, we notice that series (3) satisfies the conditions of Lemma 1 and we use the following obvious remark: if we are given two series $\sum f_n^{(l)}$ $(l=1,2)$ for which $\sum \mu\{x; f_n^{(1)} \neq f_n^{(2)}\} < \infty$, then the π-universality of one of them implies the π-universality of the other.

We now define the system ϕ that appears in the statement of the theorem. Fix numbers $k_s \uparrow$ so that

$$|\psi_s^{(j)}(x)| < 2^{k_s} \quad (j=1,2). \tag{6}$$

Renumber the elements of the system $\{\psi_n^{(0)}\}$ in the following manner: $\{\psi_s^{(j)}\}$ $(2 < j \leq 2^{k_s}; s=1,2,\ldots)$. Set

$$\phi_s^{(i)} = \sum_{j=1}^{2^{k_s}} a_{ij}^{(k_s)} \psi_s^{(j)} \quad (1 \leq i \leq 2^{k_s}),$$

where A_k are the matrices from § 1. That the system $\phi = \bigcup\limits_s \{\phi_s^{(i)}\}$ is orthonormal and closed is immediately clear. Thus

$$|\phi_s^{(i)}(x)| \leq 2^{-k_s/2}(\psi_s^{(1)} + \psi_s^{(2)}) + \sum_{j>2} |a_{ij}^{(k_s)}| < C_1,$$

because of the properties of the matrices, point (a) of Lemma 2, and inequality (6). Suppose $f = \sum\limits_s \gamma_s \psi_s^{(1)}$. Then, on the basis of (b) and (c), $f \in \bigcap\limits_{p<2} L^p$. It is easy to see that the Fourier series (1) has the form

$$f \sim \sum_{s=1}^{\infty} \sum_{i=1}^{2^{k_s}} \gamma_s 2^{-k_s/2} \phi_s^{(i)}. \tag{7}$$

Let us consider the majorants $\delta_s(f; x) = \max\limits_{1 \leq v \leq 2^{k_s}} \left| \sum\limits_{i=1}^{v} \gamma_s 2^{-k_s/2} \phi_s^{(i)}(x) \right|$. We have

$$\delta_s(f;x) = \gamma_s 2^{-k_s/2} \max_v \left| \sum_{i=1}^{v} \sum_{j=1}^{2^{k_s}} a_{ij}^{(k_s)} \psi_s^{(j)}(x) \right| \leq \gamma_s 2^{-k_s/2} \left[\sum_{j=1}^{2} \sum_{i=1}^{2^{k_s}} |a_{ij}^{(k_s)}| \, |\psi_s^{(j)}(x)| \right.$$

$$+ \max_v \sum_{j=3}^{2^{k_s}} \left| \sum_{i=1}^{l} a_{ij}^{(k_s)} \right| \Bigg] \leq \gamma_s \left[|\psi_s^{(1)}(x)| + |\psi_s^{(2)}(x)| + \max_v \sum_{j=1}^{2^{k_s}} \left| \sum_{i=1}^{v} a_{i1}^{(k_s)} a_{ij}^{(k_s)} \right| \right].$$

Property 5 §1 gives $\sum_{j} \left| \sum_{i=1}^{v} a_{i1}^{(k)} a_{ij}^{(k)} \right| \equiv l_v(1) < C$. Taking into account points (b) and (c) of Lemma 2 we conclude that $\delta_s(f;x) = o_x(1)$ almost everywhere. This relation shows that if in the series

$$\sum_s \left(\left[\sum_{i=1}^{2^{k_s-1}} 2^{-k_s/2} \gamma_s \phi_s^{(i)} \right] + \left[\sum_{i=2^{k_s-1}+1}^{2^{k_s}} 2^{-k_s/2} \gamma_s \phi_s^{(i)} \right] \right) \equiv \sum_s (\sigma_{2s-1} + \sigma_{2s})$$

the terms σ_n are arbitrarily rearranged and after that the brackets are omitted, then the latter operation does not have any influence on the convergence almost everywhere. Thus, in order to verify the π-universality of series (7) it is sufficient to verify this property for the series $\sum \sigma_n$. For this in turn it is sufficient to notice the following identities:

$$\sigma_{2s-1} = \frac{\gamma_s}{2} [\psi_s^{(1)} + \psi_s^{(2)}]; \qquad \sigma_{2s} = \frac{\gamma_s}{2} [\psi_s^{(1)} - \psi_s^{(2)}],$$

and to use point (c) and the relation $\sum |\gamma_s \psi_s^{(1)}(x)| < \infty \; (x > 0)$, which follows from (b).

We make several remarks concerning the theorem just proved.

1. Using the theorem of M. Riesz on Fourier coefficients with respect to bounded systems, we conclude that the coefficients of the series constructed satisfy the condition $\{c_n\} \in l_q \; (\forall q > 2)$; that is, they have the maximum possible rate of decrease (in the l_q scale).

2. The π-universality of any series implies, after some rearrangement, universality in the usual sense (Talalyan [137]). The latter means that for any measurable function g there can be found a sequence of partial sums S_{v_i} converging to g almost everywhere. This property, in particular, is possessed by the Fourier series constructed above.

3. By supposing $g = 0$ and ordering series (1) in an appropriate manner we obtain the following result: *there exists a nontrivial Fourier series of the class $L^p, p < 2$, with respect to a complete bounded system such that the series converges to zero almost everywhere.*

An example of a nontrivial trigonometric series converging to zero almost everywhere (a so-called null series) was, as is well known, constructed by Menshov. However this series was of course not a Fourier series. As we see, it is possible for general systems that a null series is a Fourier series.

4. Similarly, by letting $g = \infty$, we conclude that *there exists a Fourier series from $L^p, p < 2$, with respect to a complete bounded system such that the series diverges to $+\infty$ almost everywhere.*

Remember, in connection with this result, that the question of the existence of a trigonometric series diverging to ∞ almost everywhere has not been decided. Ulyanov [161] constructed an example of an orthogonol series that has such a property and that is the Fourier series in an improper sense of some function $f \notin L$. An example of an orthogonal series, with respect to a bounded system, that diverges to ∞, was first constructed by Ovsepyan and Talalyan [111]. Ovsepyan [110], using our method of A_k-transformations, investigated quite thoroughly the question of the rate of decrease of the coefficients of such series.

5. Fridlyand [33], following the scheme presented above and developing the construction of Lemma 2, showed that the system ϕ in the Theorem above can be constructed from algebraic polynomials. Moreover, the f can be any pre-assigned function that satisfies the trivially necessary condition $f \notin L^2$.

6. From the results of Arutyunyan and Gundy (Chap. III, § 5) it follows that the Fourier-Haar series of a function $f \in L$ converges unconditionally in measure to f. This implies that if $\{\phi_n\}$ is an arbitrarily ordered Haar system, and it is known that the Fourier series (1) converges almost everywhere, then relation (2) is satisfied. We do not know whether this result is true for the trigonometric system.

§ 5. Conditional Bases in Hilbert Space

In this section we examine expansions with respect to nonorthogonal bases in the space L^2. Since the question here concerns convergence in mean, the discussion will be in terms of the geometric language of Hilbert spaces. For the basic facts concerning bases in a Hilbert space H see [41].

The basis $\psi = \{\psi_k\}$ is called quasi-normal if the condition $0 < K_1 < \|\psi_k\| < K_2$ is satisfied.

If ϕ is a complete orthonormal system, and T is a linear homeomorphism of the space H onto itself, then the system $\psi_k = T \phi_k$ clearly forms a quasi-normal basis. Bases obtained in this manner are called *Riesz bases*. In other words a basis is Riesz if the space of coefficients of expansions with respect to it coincides with l_2 (N. K. Bary, see [47]).

An equivalent characterization is known: the system ψ is a Riesz basis if and only if it is quasi-normal and forms an unconditional basis in H.

For a long time it remained unknown whether or not there exists a quasi-normal basis in H which is not a Riesz basis. This problem was solved by K. I. Babenko [6], who showed that the system of functions $\psi_k^\alpha(t) = |t|^\alpha e^{ikt} \ (-\infty < k < \infty)$ for each α, $0 < |\alpha| < \frac{1}{2}$, forms such a basis in $L^2(-\pi, \pi)$. The fundamental step here consists of proving that this system is a basis, or, what is the same thing, that the trigonometric system is a basis in the space L_μ^2 with measure $d\mu = |t|^\alpha dt$. The quasi-normality of the system ψ^α is clear. Since the operator of multiplication by $|t|^\alpha \ (\alpha \neq 0)$ is not a homeomorphism in L^2, we obtained that the basis ψ^α is conditional.

We give below a purely geometric construction of conditional bases in Hilbert space [105], differing from the analytic method of Babenko. It depends on the properties of the matrices A_k (§ 1). In this way the geometric nature of conditional bases is made clear: we have in mind the properties of operators that transform an orthonormal basis into a conditional one. The spectral description of such operators [105] is contained in this section.

Construction of conditional bases. Let $\mathbb{R}^n$ be Euclidean space of dimension $n = 2^k$, and let $\{e_j\}$ $(1 \leq j \leq n)$ be an orthonormal basis in this space. Fix $\alpha, \frac{1}{\sqrt{2}} < \alpha < \sqrt{2}$ and let $\lambda_j = \alpha^{k - s(j)}, 1 \leq j \leq n$, where $s(j)$ is determined by the condition $2^{s(j)} < j \leq 2^{s(j)+1}$, $s(1) = 0$. The equalities

$$T_k e_j = \lambda_j e_j \quad (1 \leq j \leq n) \tag{1}$$

determine a linear operator $T_k : \mathbb{R}^n \to \mathbb{R}^n$.

We pass by means of the transformation A_k to a new orthonormal basis $\phi_i = \sum_{j=1}^{n} a_{ij}^{(k)} e_j$ and designate $\psi_i = T_k \phi_i$. As a consequence of relation (1) § 1, we have

$$\|\psi_i\|^2 = \sum_j |\lambda_j a_{ij}^{(k)}|^2 = \alpha^{2k} 2^{-k} + \sum_{s=0}^{k-1} \alpha^{2(k-s)} \sum_{j=2^s+1}^{2^{s+1}} |a_{ij}^{(k)}|^2 = \sum_{s=0}^{k} \left(\frac{\alpha^2}{2}\right)^{k-s}.$$

Hence,

$$0 < C_2(\alpha) < \|\psi_i\| < C_1(\alpha) \quad (\forall i), \tag{2}$$

where C_1, C_2 are constants independent of k.

Clearly the collection of vectors $\psi = \psi(k, \alpha) = \{\psi_i\}$ $(1 \leq i \leq n)$ forms a basis in $\mathbb{R}^n$; that is, to each $x \in \mathbb{R}^n$ there corresponds a unique expansion $x = \sum \alpha_i \psi_i$. We estimate the "Banach constant" $K(\psi) = \max_v \|Q_v\|$, where Q_v are the projections corresponding to this expansion: $Q_v(x) = \sum_{i=1}^{v} \alpha_i \psi_i$. For this we introduce in $\mathbb{R}^n$, in addition to its Euclidean metric, the norms of l_1^n and l_∞^n; that is, we define for each $x = \sum c_j e_j$.

$$\|x\|' = \sum_j |c_j|, \qquad \|x\|'' = \max_j |c_j|.$$

Clearly, $e_j = \frac{1}{\lambda_j} \sum_i a_{ij}^{(k)} \psi_i$. Using the notation of § 1, we have

$$Q_v e_j = \frac{1}{\lambda_j} \sum_{i=1}^{v} a_{ij}^{(k)} \psi_i = \frac{1}{\lambda_j} \sum_{i=1}^{v} a_{ij}^{(k)} \sum_{r=1}^{n} a_{ir}^{(k)} \lambda_r e_r = \frac{1}{\lambda_j} \sum_{r=1}^{n} d_v(j,r) \lambda_r e_r.$$

Property (iv) § 1 gives

$$\|Q_v x\|' = \|Q_v(\sum c_j e_j)\|' \leq \sum |c_j| \|Q_v e_j\|'$$

$$\leq \|x\|' \max_v \frac{1}{\lambda_j} \sum_r |d_v(j,r)| \lambda_r \leq C(\alpha) \|x\|'.$$

Analogously,

$$\|Q_v x\|'' = \left\| \sum_j \frac{c_j}{\lambda_j} \sum_r d_v(j,r) \lambda_r e_r \right\|'' = \max_r \left| \sum_j \frac{c_j}{\lambda_j} d_v(j,r) \lambda_r \right|$$

$$\leq \|x\|'' \max_r \sum_j |d_v(j,r)| \frac{\lambda_r}{\lambda_j} = \|x\|'' \max_r \sum_j |d_v(r,j)| \frac{\lambda_r}{\lambda_j}$$

$$= \|x\|'' \max_j \sum_r |d_v(j,r)| \frac{\lambda_j}{\lambda_r} \leq C(\alpha) \|x\|''$$

(the same property is used for $\alpha' = \frac{1}{\alpha}$).

After interpolating we obtain

$$K(\psi) \leq C(\alpha). \tag{3}$$

At the same time, the inequality

$$L(\psi) \equiv \|T_k\| + \|T_k^{-1}\| = \max \lambda_j + \max \lambda_j^{-1} > \begin{cases} \alpha^k, \alpha > 1 \\ \alpha^{-k}, \alpha < 1 \end{cases} \tag{4}$$

is fulfilled. Thus, if $\alpha \neq 1$ is fixed, and k increases, then inequalities (2) and (3) are satisfied uniformly, and the constants $L(\psi)$ increase to infinity. This means that *the system $\psi(k, \alpha)$ is a finite-dimensional model of a quasi-normal non-Riesz basis*.

By now it is clear that if a Hilbert space H is decomposed as a direct sum of pairwise orthogonal subspaces $\mathbb{R}^n, n = 2^k, k = 1, 2, \ldots,$ and in each of these the construction mentioned above is done for a fixed $\alpha \neq 1$, then *the system $\psi^\alpha = \bigcup_k \psi(k, \alpha)$ forms a quasi-normal basis in H that is not a Riesz basis (that is, it is a conditional basis)*.

Observation. Inequality (4) shows that the "constant of unconditionality" of the bases $\psi(k, \alpha)$ as $k \to \infty$ has the order of growth $n^{1/2 - \varepsilon}$, where $\varepsilon(\alpha) > 0$ is arbitrarily small if α is sufficiently close to the boundary of the interval $\left(\dfrac{1}{\sqrt{2}}, \sqrt{2}\right)$. The Babenko bases possess this same property (see [166]). It is interesting that in this sense both these families of bases are extremal: *for any C there exists an $\varepsilon(C) > 0$ that satisfies the relation*

$$\sup L(\psi) = O(n^{1/2 - \varepsilon}),$$

where the sup is taken over all normed bases ψ in $\mathbb{R}^n$ with constant $K(\psi) < C$.

This result [105] follows easily from the following general theorem, which is proved in the work [45].

Let $\{\psi_i\}$ be an arbitrary normed basis in the uniformly convex uniformly smooth space B (in particular, in H). Then there exist positive numbers $A_1, A_2, p_1 > 1, p_2 < \infty$ (depending on ψ) such that for any $x = \sum c_i \psi_i \in B$ the inequality

$$A_2 \|c\|_{l_{p_2}} \leq \|x\|_B \leq A_1 \|c\|_{l_{p_1}}$$

is satisfied.

The Spectral Description of Generative Operators. We call a bounded linear operator $A: H \to H$ *generative* if there exists an orthonormal system of vectors $\{\phi_n\}$ such that the system $\psi_n = A\phi_n$ forms a quasi-normal basis in H. For example, the operator of multiplication by $|t|^\alpha, 0 \leq \alpha < 1/2$, in $L^2(-\pi, \pi)$ is generative, according to the theorem of Babenko.

We propose a problem: to describe all generative operators. It is easy to reduce this to the case where A is a positive operator. Indeed, in order that the operator A be generative it is necessary that it satisfy the condition

1. The point $\lambda = 0$ is not an eigenvalue either for A or for the dual operator A^*.

This condition ensures that the relation $A = UA_1$ is true, where $A_1 = (A^*A)^{1/2}$, and U is some unitary operator. Hence it follows that A_1 is generative whenever A is.

Let the operator A be positive. The case where the point $\lambda = 0$ does not lie in the spectrum is trivial. In this case A represents a homeomorphism of the space H onto itself, and consequently is generative. The only interesting case is when

(i) *the point λ lies in the spectrum but is not in the discrete spectrum.*

The following proposition [105] holds.

Theorem. *In order for a positive linear operator A satisfying condition* (i) *to be generative the following condition is necessary and sufficient:*

(ii) *there exists a number $0 < q < 1$ such that each interval $[q^{n+1}, q^n]$ $(n = 1, 2, \ldots)$ in the spectral decomposition of the operator A corresponds to an infinite-dimensional invariant subspace.*

Roughly speaking, the spectrum of a generative operator cannot have lacunae too large in a neighborhood of zero.

The sufficiency of the condition of the theorem follows easily from the construction demonstrated in the preceding subsection. The generative operator $T^\alpha = \bigoplus T_n^\alpha$ of the basis ψ^α constructed there has a point spectrum: the eigenvalues are equal to α^k and each of them has infinite multiplicity. Suppose we are given an operator A, $\|A\| < 1$, satisfying conditions (i) and (ii). The corresponding decomposition of the identity (see [113]) is denoted by E_λ. For any interval Δ let $H(\Delta) = E(\Delta)(H) = \left(\int_\Delta dE_\lambda\right)(H)$. It is easy to select a number b so that each half-open interval $\Delta_n = (b^{n+1}, b^n]$ contains at least one term of the sequence $\mu_k = \{\alpha^k\}$, and $\dim H_n = \infty$, $H_n = H(\Delta_n)$. Let $\mu_n^{(l)}$ $(1 \le l \le l_n)$ be all the terms of the sequence referred to that belong to the half-open interval Δ_n. We express H_n as a direct sum of pairwise orthogonal infinite-dimensional invariant subspaces $H_n = \bigoplus_{1 \le l \le l_n} H_n^l$, and set

$$\tilde{T}x = \mu_n^{(l)} x \qquad (x \in H_n^l).$$

This defines in H_n^l, and also in the whole space $H = \bigoplus H_n$, a linear operator $\tilde{T}$ that is unitarily equivalent to the operator T^α. At the same time it follows easily from the construction that $A = S\tilde{T}$, where S is some linear homeomorphism of the space H onto itself. This proves that the operator A is generative.

We turn to the proof of the necessity of the condition of the theorem. Suppose we are given an operator A, $\|A\| < 1$, and an orthonormal basis $\{\phi_n\}$ such that the system $\psi_n = A\phi_n$ satisfies the condition $\|\psi_n\| \ge \gamma > 0$ and forms a basis in H. Denote the dual system of functionals by $\{d_n(x)\}$. We introduce into consideration the projections $Q_\nu x = \sum_1^\nu d_k(x)\phi_k$ and the Banach constant $K = \sup_\nu \|Q_\nu\|$. Denote by ϕ_s^m (ϕ_s^∞) the subspace generated by the vectors $\{\phi_i\}$ $(s \le i \le m)$ (respectively $s \le i < \infty$).

$P(L)$ will denote the orthogonal projection onto the subspace $L \subset H$. We choose a number q satisfying the inequality

$$0 < q < \frac{\gamma}{32K}. \tag{5}$$

Fix a natural number n and suppose

$$L = H((0, q^{n+1}]), \quad L' = H((q^{n+1}, q^n]), \quad L'' = H((q^n, 1]).$$

Because of condition (i) we have

$$L \oplus L' \oplus L'' = H, \quad \dim L = \infty. \tag{6}$$

Clearly, the subspaces L, L', L'' are invariant under the operator A, and

$$\|A\|_L \le q^{n+1}, \quad \|Ax\| \ge q^n \|x\| \quad (x \in L'') \tag{7}$$

($\|A\|_L$ denotes the norm of the restriction of the operator A to the subspace L).

We shall show that $\dim L' = \infty$. Otherwise we can find a number s so that the condition

$$\|P(\phi_s^x)\|_{L'} < \tfrac{1}{8} \tag{8}$$

is satisfied. For any $x \in L$ we have, from (7),

$$x = \sum_n (x, \phi_k) \phi_k; \quad Ax = \sum (x, \phi_k) \psi_k \equiv \sum d_k(Ax) \psi_k;$$

$$|\gamma(x, \phi_k)| \le \|(x, \phi_k)\psi_k\| = \|Q_k(Ax) - Q_{k-1}(Ax)\|$$

$$\le 2 K q^{n+1} \|x\|.$$

Hence because of (5) it follows that $|(x, \phi_k)| \le \tfrac{1}{8} \|x\|$; that is,

$$\|P(\phi_k^k)\|_L \le \tfrac{1}{8} \quad (\forall k). \tag{9}$$

At the same time

$$\lim_{k \to x} \|P(\phi_s^k)\|_L = 1 \tag{10}$$

(it is sufficient to notice that because of the inequality $\infty = \dim L > s - 1$ we can find a vector $x \in L \cap \phi_s^x, x \ne 0$). Since $P(\phi_s^k) = P(\phi_s^{k-1}) + P(\phi_k^k), k > s$, we find from (8) and (9) a number m such that $3/8 \le \|P(\phi_s^m)\| < 5/8$. Hence because of (6) and (8) it follows that

$$\|P(L'')P(\phi_s^m)\|_L = \|(I - P(L) - P(L'))P(\phi_s^m)\|_L$$

$$\ge \|P(\phi_s^m)\|_L [1 - \|P(\phi_s^m)\|_L - \|P(\phi_s^m)\|_{L'}] > \tfrac{1}{16}$$

(in view of the obvious equality $\|P(\phi)\|_M = \|P(M)\|_\phi$).

Let $x \in L, \|x\| = 1, \|P(L'')P(\phi_s^m)x\| > 1/16$. Then $P(\phi_s^m)x = x' + x''$, where $x'' \in L'', \|x''\| > 1/16, x' \perp L''$. Clearly, $Ax = \sum (x, \phi_k)\psi_k$. Because of (7) we have

$$q^{n+1} \ge \|Ax\| > \frac{1}{2K} \left\| \sum_{k=s}^m (x, \phi_k)\psi_k \right\| = \frac{1}{2K} \|AP(\phi_s^m)x\|$$

$$= \frac{1}{2K} \|Ax' + Ax''\| \ge \frac{1}{2K} \|Ax''\| \ge \frac{q^n \gamma}{32 K},$$

which contradicts (5). Thus, the space L' is infinite-dimensional for any n; that is, condition (ii) is fulfilled. The proof of the theorem is now complete.

If A is an arbitrary bounded (not necessarily positive) operator, then because of the discussion above, *it is generative if and only if condition* (i) *is satisfied and* *the spectrum of the operator* $(AA^*)^{1/2}$ *either does not contain zero or satisfies condition* (ii).

In the latter case the quasi-normal basis produced by this operator is conditional.

For some concrete consequences of this theorem see [105].

We note that the theorem proved does not resolve the following problem: for a given ON basis, which operators transform it into another basis? In particular, Gaposhkin [35], generalizing the theorem of Babenko, found sufficient conditions on the function $\lambda(t)$ for the system $\psi(t) = \lambda(t)e^{int}$ to form a basis in $L^2(-\pi, \pi)$. It turns out that these conditions are close to being necessary (see [52]). The analogous question for the Haar system is treated in § 3, Chap. III.

Bibliography

1. Alexits, G.: Konvergenzprobleme der Orthogonalreihen. Budapest: Verlag der Ungarischen Akademie der Wissenschaften 1960. Russian translation: Moscow, 1963.
2. Alexits, G., Sharma, P.: The influence of Lebesgue functions in the convergence and summability of function series. Acta Sci. Math. **33**, 1–10 (1972).
3. Arutyunyan, F.G.: The convergence almost everywhere of series with respect to bases in the space $L^p[0,1]$. Dokl. Akad. Nauk Arm. SSR **38**, No. 3, 129–134 (1964) [Russian].
4. Arutyunyan, F.G.: Series with respect to the Haar system. Dokl. Akad. Nauk Arm. SSR **42**, No. 3, 134–140 (1966) [Russian].
4′. Arutyunyan, F. G.: Representation of measurable functions by series convergent almost everywhere. Mat. Sb. **90**, 483–520 (1973) [Russian].
5. Arutyunyan, F.G.: Bases of the spaces $L[0,1]$ and $C[0,1]$. Mat. Zametki **11**, 241–250 (1972) [Russian]. = Math. Notes **11**, 152–157 (1972).
6. Babenko, K.I.: On conjugate functions. Dokl. Akad. Nauk SSSR **62**, 157–160 (1948) [Russian].
7. Balashov, L.A., Rubinšteĭn, A.I.: Series with respect to the Walsh system and their generalizations. In the collection Itogi Nauki: Matem. Analiz. Moscow, 1971, 147–202 [Russian].
8. Banach, S.: Théorie des Opérations Linéaires. New York: Chelsea 1955. Russian translation: Kiev, 1948.
9. Banach, S.: Sur la divergence des séries orthogonales. Studia Math. **9**, 139–154 (1940).
10. Bary, N.K.: Trigonometric series. Moscow, 1961. English translation: New York: Pergamon 1964.
11. Billard, P.: Sur la convergence presque partout des séries de Fourier-Walsh des fonctions de l'espace $L^2[0,1]$. Studia Math. **28**, 363–388 (1966–67).
12. Boas, R.: Integrability theorems for trigonometric transforms. Berlin-New York: Springer-Verlag 1967.
13. Bochkarev, S.V.: The unconditional convergence almost everywhere of Fourier-Haar series of continuous functions. Mat. Zametki **4**, 211–220 (1968) [Russian]. = Math. Notes **4**, 618–623 (1968).
14. Bochkarev, S.V.: The Fourier coefficients of a function of the class Lip α with respect to complete orthonormal systems. Mat. Zametki **7**, 397–402 (1970) [Russian]. = Math. Notes **7**, 239–242 (1970).
15. Bochkarev, S.V.: Absolute convergence of Fourier series with respect to complete orthonormal systems. Uspehi Mat. Nauk **27**, 53–76 (1972) [Russian].
15′. Bochkarev, S. V.: On the absolute convergence of Fourier series in complete bounded orthonormal systems of functions. Mat. Sb. **93**, 203–217 (1974) [Russian].
16. Bochkarev, S. V.: Unconditional bases. Mat. Zametki **1**, 391–398 (1967) [Russian]. = Math. Notes **1**, 261–265 (1967).
17. Bürkholder, D., Gundy, R.: Extrapolation and interpolation of quasi-linear operators on martingales. Acta Math. **124**, 249–304 (1970).
18. Carleson, L.: On convergence and growth of partial sums of Fourier series. Acta Math. **116**, 135–157 (1966).
19. Chanturiya, Z.A.: The stability of the orthogonalization process and its applications. Studia Math. **41**, 273–290 (1972) [Russian].

20. Ciesielski, Z.: Properties of the orthonormal Franklin system. Studia Math. **23**, 141–157 (1963); **27**, 289–323 (1966).
21. Ciesielski, Z.: Construction of an orthonormal basis in $C^M(I^z)$. Trudy Meždunarodnoĭ Konferencii po teorii funkciĭ, Sofia, 1972, 147–150.
22. Ciesielski, Z., Musielak, J.: On absolute convergence of Haar series. Colloq. Math. **7**, 61–65 (1959).
23. Cohen, P.: On a conjecture of Littlewood and idempotent measures. Amer. J. Math. **82**, 191–212 (1960).
24. Davenport, H.: On a theorem of P. Cohen. Mathematika **7**, 93–97 (1960).
25. Davis, B.: On the integrability of the martingale square function. Israel J. Math. **8**, 187–190 (1970).
26. Day, M.: Normed linear spaces. Berlin, Heidelberg, New York: Springer 1973.
27. Dunford, N., Schwartz, J.: Linear Operators, v.1. New York: Interscience 1958. Russian translation: Moscow, 1962.
28. Efimov, A.V.: On orthogonal series not summable by linear methods. Dokl. Akad. Nauk SSSR **152**, 31–34 (1963) [Russian]. = Soviet Math. Dokl. **4**, 1227–1230 (1963).
29. Efimov, A.V.: On the summation of orthogonal series with the de la Vallée-Poussin means. Izv. Akad. Nauk SSSR Ser. Mat. **27**, 831–842 (1963) [Russian].
30. Efimov, A.V.: A generalization of a theorem of Kaczmarz. Mat. Zametki **1**, 399–404 (1967) [Russian]. = Math. Notes **1**, 266–269 (1967).
31. Enflo, P.: A counterexample to the approximation problem. Acta Math. **130**, 309–317 (1973).
32. Fefferman, C.: The multiplier problem for the ball. Ann. of Math. **94**, 330–336 (1971).
33. Fridlyand, Yu. S.: On a π-universal Fourier series. Moskov. Inst. Èlektron. Mašino-stroenija—Trudy MIÈM **15**, 3–15 (1971) [Russian]. Izv. Akad. Nauk Armjan. SSR Ser. Mat. **8**, 329–344 (1973) [Russian].
34. Fridlyand, Yu. S.: On an exact Weyl multiplier for complete orthonormal systems bounded in the uniform norm. Moskov. Inst. Èlektron. Mašinostroenija—Trudy MIÈM **24**, 22–31 (1972) [Russian].
34'. Fridlyand, Yu. S.: An irremovable Carleman singularity for Haar's system. Mat. Zametki **14**, 799–807 (1973) [Russian]. = Math. Notes **14**, 1017–1022 (1973).
35. Gaposhkin, V.F.: A generalization of a theorem of M. Riesz on conjugate functions. Mat. Sb. **46**, 359–372 (1958) [Russian].
36. Gaposhkin, V.F.: Unconditional bases in L^p-spaces. Uspehi Mat. Nauk **13**, 179–184 (1958) [Russian].
37. Gaposhkin, V.F.: On the convergence of orthogonal series. Dokl. Akad. Nauk SSSR **159**, 243–246 (1964) [Russian]. = Soviet Math. Dokl. **5**, 1445–1448 (1964).
38. Gaposhkin, V.F.: On the existence of unconditional bases in Orlicz spaces. Funkcional. Anal. i Priložen. **1**, No. 4, 26–32 (1967) [Russian].
39. Gaposhkin, V.F.: Lacunary series and independent functions. Uspehi Mat. Nauk **21**, No. 6, 3–82 (1966) [Russian].
40. Garsia, A.: Existence of almost everywhere convergent rearrangements for Fourier series of L^2 functions. Ann. of Math. **79**, 623–629 (1964).
41. Gokhberg, I.Ts., Kreĭn, M.G.: Introduction to the theory of linear nonselfadjoint operators. Moscow, 1965 [Russian].
42. Golubov, B.I.: Fourier series of continuous functions with respect to the Haar system. Izv. Akad. Nauk SSSR Ser. Mat. **28**, 1271–1296 (1964) [Russian].
43. Golubov, B.I.: Series with respect to the Haar system. In the collection Itogi Nauki: Matem. Analiz. Moscow, 1971, 109–146 [Russian].
44. Gundy, R.: Martingale theory and pointwise convergence of certain orthogonal series. Trans. Amer. Math. Soc. **124**, 228–248 (1966).
45. Gurariĭ, V.I., Gurariĭ, N.I.: Bases in uniformly convex and uniformly smooth Banach spaces. Izv. Akad. Nauk SSSR Ser. Mat. **35**, 210–215 (1971) [Russian]. = Math. USSR—Izv. **5**, 220–225 (1971).
46. Gurariĭ, V.I., Meletidi, M.A.: Stability of completeness of sequences in Banach spaces. Bull. Acad. Polon. Sci. **18**, 533–536 (1970) [Russian].

47. Guter, R.S., Ulyanov, P.L.: On new results in the theory of orthogonal series. 333–456. In the appendix to the Russian translation of [55]. [Russian].
48. Haar, A.: Zur Theorie der orthogonalen Funktionensysteme. Math. Ann. **69**, 331–371 (1910).
49. Haar, A.: Über einige Eigenschaften der Orthogonalen Funktionensysteme. Math. Z. **31**, 128–137 (1929).
50. Halmos, P.: Measure Theory. New York: Van Nostrand 1950. Russian translation: Moscow, 1953.
51. Hardy, G., Littlewood, J.: A new proof of a theorem on rearrangements. J. London Math. Soc. **23**, 163–168 (1948).
52. Helson, H., Szegö, G.: A problem in prediction theory. Ann. Mat. Pura Appl. **51**, 109–138 (1960).
53. Hunt, R.: On the convergence of Fourier series. Orthogonal expansions and their continuous analogues. Southern Illinois University Press, 235–256, 1968.
54. Il'in, V.A.: Problems of localization and convergence of Fourier series with respect to fundamental systems of functions of the Laplace operator. Uspehi Mat. Nauk **23**, No. 2, 60–120 (1968) [Russian].
55. Kaczmarz, S., Steinhaus, H.: Theorie der Orthogonalreihen. Warsaw: Monografje Matematyczne 1935. Russian translation: Moscow, 1958.
56. Kahane, J.-P.: Séries de Fourier absolument convergentes. Berlin-New York: Springer-Verlag 1970.
57. Kahane, J.-P.: Some random series of functions. 1968, Massachusetts.
58. Kahane, J.-P.: Sur les réarrangements des suites de coefficients de Fourier-Lebesgue. C.R. Acad. Sci. Paris **265**, 310–312 (1967).
59. Kolmogorov, A.: Une série de Fourier-Lebesgue divergente partout. C.R. Acad. Sci. Paris **183**, 1327–1328 (1926).
60. Kolmogorov, A.: Une contribution à l'étude de convergence des séries de Fourier. Fund. Math. **5**, 96–97 (1924).
61. Kolmogorov, A., Menshov, D.: Sur la convergence des séries de fonctions orthogonales. Math. Z. **26**, 432–441 (1927).
62. Kozlov, V.Ya.: On a local characteristic of a complete orthogonal normalized system of functions. Math. Sb. **23**, 441–474 (1948) [Russian].
63. Krantsberg, A.S.: The order of growth of orthonormal functions with a bounded number of changes of sign. Moskov. Inst. Èlektron. Mašinostroenija—Trudy MIÈM **4**, 69–78 (1968) [Russian].
63'. Krantsberg, A.S.: Rearrangements of the Haar system. Mat. Zametki **15**, 63–71 (1974) [Russian]. = Math. Notes **15**, 35–39 (1974).
64. Krantsberg, A.S.: On whether the Haar system is a basis in weighted spaces. Moskov. Inst. Èlektron. Mašinostroenija—Trudy MIÈM **24**, 14–21 (1972) [Russian].
64'. Krantsberg, A.S.: On divergent orthogonal Fourier series. Mat. Sb. **93**, 540–553 (1974) [Russian].
65. Krasnoselskiĭ, M.A., Rutitskiĭ, Ya.B.: Convex functions and Orlicz spaces. Moscow, 1958 [Russian]. English translation: New York: Gordon and Breach 1961.
66. Kuptsov, N.P.: Localization of equiconvergence theorems. Mat. Sb. **74**, 554–564 (1967) [Russian]. = Math. USSR—Sb. **3**, 509–518 (1967).
67. Leindler, L.: Über die orthogonalen Polynomsysteme. Acta Sci. Math. **21**, 19–46 (1960).
68. Leindler, L.: Nicht verbesserbare Summierbarkeitsbedingungen für Orthogonalreihen. Acta Math. Acad. Sci. Hungar. **13**, 425–432 (1962).
69. Leindler, L.: Über die unbedingte Konvergenz der Orthogonalreihen. Publ. Math. Debrecen **11**, 139–148 (1964).
70. Leindler, L.: Abschätzungen für die Partialsummen und für die $(R, \lambda(n), 1)$-Mittel allgemeiner Orthogonalreihen. Acta Sci. Math. **23**, 227–236 (1962).
71. Makhmudov, A.S.: On Fourier and Taylor coefficients of continuous functions. I Izv. Akad. Nauk Azerbaĭdžan. SSR, No. 2, 23–29 (1964); II ibid., No. 4, 35–44 (1964); III In the collection "Some questions of functional analysis and its application", 103–117, Baku, 1965 [Russian].

72. Marcinkiewicz, J.: Quelques théorèmes sur les séries orthogonales. Ann. Soc. Polon. Math. **16**, 85–96 (1937).

73. Meletidi, M.A.: Bases in the spaces C and L^p. Mat. Zametki **10**, 635–640 (1971) [Russian]. = Math. Notes **10**, 812–815 (1971).

74. Menshov, D.: Sur les séries de fonctions orthogonales. Fund. Math. **4**, 82–105 (1923); **8**, 56–108 (1926); **10**, 375–420 (1927).

75. Menshov, D.: Sur les multiplicateurs de convergence pour les séries de polynomes orthogonaux. Mat. Sb. **6**, 27–52 (1939).

76. Menshov, D.: On the summation of orthogonal series by linear methods. Trudy Moskov. Mat. Obšč. **10**, 351–418 (1961) [Russian].

77. Milman, V.D.: Geometric theory of Banach space, I. Theory of basic and minimal systems. Uspehi Mat. Nauk **25**, No. 3, 113–174 (1970) [Russian].

78. Mityagin, B.S.: On the absolute convergence of the series of Fourier coefficients. Dokl. Akad. Nauk SSSR **157**, 1047–1050 (1964) [Russian]. = Soviet Math. Dokl. **5**, 1083–1086 (1964).

79. Móricz, F.: On the order of magnitude of the partial sums of rearranged Fourier series of square integrable functions. Acta Sci. Math. **28**, 155–167 (1967).

79'. Móricz, F.: On unconditional convergence of series in the Haar system. Izv. Akad. Nauk SSSR Ser. Mat. **22**, 1229–1238 (1963) [Russian].

79''. Móricz, F.: The order of magnitude of the Lebesgue functions and summability of function series. Acta Sci. Math. **34**, 289–296 (1973).

80. Móricz, F., Tandori, K.: On a problem of summability of orthogonal series. Acta Sci. Math. **29**, 331–350 (1968).

81. Mushegyan, G.M., Ovsepyan, R.I.: The uniqueness of orthogonal series. Izv. Akad. Nauk Armjan. SSR Ser. Mat. **4**, 259–266 (1969) [Russian].

82. Nakata, S.: On the divergence of rearranged Fourier series of square integrable functions. Acta Sci. Math. **32**, 59–70 (1971).

83. Nikishin, E.M.: Convergence of certain series of functions. Izv. Akad. Nauk SSSR Ser. Mat. **31**, 15–26 (1967) [Russian]. = Math. USSR—Izv. **1**, 13–24 (1967).

84. Nikishin, E.M.: Resonance theorems and superlinear operators. Uspehi Mat. Nauk **25**, No. 6, 129–191 (1970) [Russian].

85. Nikishin, E.M.: Rearrangements of function series. Mat. Sb. **85**, 272–285 (1971) [Russian]. = Math. USSR—Sb. **14**, 267–280 (1971).

86. Nikishin, E.M.: A resonance theorem and series in eigenfunctions of the Laplace operator. Izv. Akad. Nauk SSSR Ser. Mat. **36**, 795–813 (1972) [Russian]. = Math. USSR—Izv. **6**, 788–806 (1972).

87. Nikishin, E.M.: Weyl multipliers for multiple Fourier series. Mat. Sb. **89**, 340–348 (1972) [Russian]. = Math. USSR—Sb. **18**, 351–360 (1972).

88. Nikishin, E.M., Ulyanov, P.L.: On absolute and unconditional convergence. Uspehi Mat. Nauk **22**, No. 3, 240–242 (1967) [Russian].

89. Nikolskiĭ, S. M.: Approximation of functions of several variables and imbedding theorems. Moscow, 1969 [Russian].

90. Olevskiĭ, A.M.: Divergent series in L^2 with respect to complete systems. Dokl. Akad. Nauk SSSR **138**, 545–548 (1961) [Russian]. = Soviet Math. Dokl. **2**, 669–672 (1961).

91. Olevskiĭ, A.M.: Divergent Fourier series of continuous functions. Dokl. Akad. Nauk SSSR **141**, 28–31 (1961) [Russian]. = Soviet Math. Dokl. **2**, 1382–1386 (1961).

92. Olevskiĭ, A.M.: On orthogonal series with respect to complete systems. Mat. Sb. **58**, 707–748 (1962) [Russian].

93. Olevskiĭ, A.M.: Divergent Fourier series. Izv. Akad. Nauk SSSR Ser. Mat. **27**, 343–366 (1963) [Russian].

94. Olevskiĭ, A.M.: Divergence of orthogonal series and Fourier coefficients of continuous functions. Sibirsk. Mat. Ž. **4**, 647–656 (1963) [Russian].

95. Olevskiĭ, A. M.: Unconditional summability of general functional and orthogonal series. Sibirsk. Mat. Ž. **5**, 1071–1097 (1964) [Russian].

96. Olevskiĭ, A.M.: On a problem of P.L. Ulyanov. Uspehi Mat. Nauk **20**, No. 2, 197–202 (1965) [Russian].

97. Olevskiĭ, A.M.: Fourier series of continuous functions with respect to bounded orthonormal systems. Izv. Akad. Nauk SSSR Ser. Mat. 30, 387–432 (1966) [Russian].
98. Olevskiĭ, A.M.: A certain orthonormal system and its applications. Mat. Sb. 71,
297–335 (1966) [Russian].
99. Olevskiĭ, A.M.: On singularities of Carleman type. Sibirsk. Mat. Ž. 8, 807–826 (1967)
[Russian]. = Siberian Math. J. 8, 611–625 (1967).
100. Olevskiĭ, A.M.: Fourier series and Lebesgue functions. Uspehi Mat. Nauk 22, No. 3,
237–239 (1967) [Russian].
101. Olevskiĭ, A.M.: On the order of growth of the Lebesgue functions of bounded orthonormal systems. Dokl. Akad. Nauk SSSR 176, 1247–1250 (1967) [Russian]. = Soviet
Math. Dokl. 8, 1311–1314 (1967).
102. Olevskiĭ, A.M.: On some peculiarities of Fourier series in L^p, $p<2$. Mat. Sb. 77,
251–258 (1968) [Russian]. = Math. USSR—Sb. 6, 233–239 (1968).
103. Olevskiĭ, A.M.: The extension of a sequence of functions to a complete orthonormal
system. Mat. Zametki 6, 737–747 (1969) [Russian]. = Math. Notes 6, 908–913 (1969).
104. Olevskiĭ, A.M.: The stability of the Schmidt orthogonalization operator. Izv. Akad.
Nauk SSSR Ser. Mat. 34, 803–826 (1970) [Russian]. = Math. USSR—Izv. 4, 811–834
(1970).
105. Olevskiĭ, A.M.: On operators generating conditional bases in a Hilbert space. Mat.
Zametki 12, 73–84 (1972) [Russian]. = Math. Notes 12, 476–482 (1972).
106. Olevskiĭ, A.M.: The localization of Carleman singularities on compact sets of measure
zero. Dokl. Akad. Nauk SSSR 202, 30–33 (1972) [Russian]. = Soviet Math. Dokl. 13,
27–30 (1972).
107. Olevskiĭ, A.M.: An example of an orthonormal system of convergence in C but not
in L^2. Mat. Sb. 91, 134–141 (1973) [Russian]. = Math. USSR—Sb. 20, 145–154 (1973).
108. Orlicz, W.: Beiträge zur Theorie der Orthogonalentwicklungen. I Studia Math. 1,
1–39 (1929); II ibid., 241–255; III Bull. Acad. Polon. Sci., 229–238 (1932); IV Studia
Math. 5, 1–14 (1934); V ibid. 6, 20–38 (1936); VI ibid. 8, 141–147 (1939).
109. Orlicz, W.: Über unbedingte Konvergenz in Funktionenräumen. Studia Math. 1,
83–85 (1930).
110. Ovsepyan, R.I.: The convergence of orthogonal series to $+\infty$. Mat. Zametki 11,
499–508 (1972) [Russian]. = Math. Notes 11, 305–310 (1972).
111. Ovsepyan, R.I., Talalyan, A.A.: The convergence of orthogonal series to $+\infty$. Mat.
Zametki 8, 129–135 (1970) [Russian]. = Math. Notes 8, 545–549 (1970).
112. Pełczyński, A.: On the impossibility of embedding the space L in certain Banach spaces.
Colloq. Math. 8, 199–203 (1961).
113. Riesz, F., Sz.-Nagy, B.: Functional analysis. New York: Ungar 1955. Russian translation: Moscow, 1954.
113′. Rjazanov, B.V., Slepchenko, A.N.: Orthogonal bases in L^p. Izv. Akad. Nauk SSSR
Ser. Mat. 34, 1159–1172 (1970) [Russian]. = Math. USSR—Izv. 4, 1169–1182 (1970).
114. Rubinšteĭn, A.I.: Rearrangements of complete systems of convergence. Sibirsk. Mat.
Ž. 13, 420–428 (1972) [Russian]. = Siberian Math. J. 13, 291–297 (1972).
115. Saks, S.: Theory of the integral. 2nd rev. ed. New York: G.E. Stechert 1937. Russian
translation: Moscow, 1949.
116. Salem, R.: On a problem of Littlewood. Amer. J. Math. 77, 535–540 (1955).
117. Sawyer, S.: Maximal inequalities of weak type. Ann. of Math. 84, 157–173 (1966).
118. Semenov, E.M.: A method for establishing interpolation theorems in symmetric spaces.
Dokl. Akad. Nauk SSSR 185, 1243–1246 (1969) [Russian]. = Soviet Math. Dokl. 10,
507–511 (1969).
119. Shaĭdukov, K.M.: The existence of an orthonormal basis in the class of polynomials.
Naučn. Trudy Kazanskogo In-ta Inž. Stroit. Neft. Prom. 5, 119–151 (1957) [Russian].
120. Shapiro, H.: Incomplete orthogonal families and related questions on orthogonal
matrices. Michigan Math. J. 11, 15–18 (1964).
121. Sidon, S.: Über orthogonale Entwicklungen. Acta Sci. Math. 10, 206–253 (1943).
122. Skvortsov, V.A.: Differentiation with respect to nets, and the Haar series. Mat. Zametki 4, 33–40 (1968) [Russian]. = Math. Notes 4, 509–513 (1968).

123. Slepchenko, A. N.: Orthogonal bases in L. Mat. Zametki **6**, 749–758 (1969) [Russian]. = Math. Notes **6**, 914–919 (1969).

124. Sobol, I. M.: Multidimensional quadrature formulas and the Haar functions. Moscow, 1969 [Russian].

125. Spitzer, F.: A combinatorial lemma and its application to probability theory. Trans. Amer. Math. Soc. **82**, 323–339 (1956).

126. Stechkin, S. B.: On Fourier coefficients of continuous functions. Izv. Akad. Nauk SSSR Ser. Mat. **21**, 93–116 (1957) [Russian].

127. Stechkin, S. B., Ulyanov, P. L.: Subsequences of convergence of series. Trudy MIAN **86**, 1–82 (1965) [Russian].

128. Stein, E.: On limits of sequences of operators. Ann. of Math. **74**, 140–170 (1961).

129. Stein, E.: Note on the class $L \log L$. Studia Math. **32**, 305–310 (1969).

130. Sunouchi, G.: On the Riesz summability of Fourier series. Tôhoku Math. J. **11**, 319–326 (1959).

131. Szlenk, W.: Une remarque sur l'ortogonalisation des bases de Schauder dans l'espace C. Colloq. Math. **15**, 297–301 (1966).

132. Sz.-Nagy, B.: Approximation properties of orthogonal expansions. Acta Sci. Math. **15**, 31–37 (1953–54).

133. Taĭkov, L. V.: The divergence of Fourier series of continuous functions with respect to a rearranged trigonometric system. Dokl. Akad. Nauk SSSR **150**, 262–265 (1963) [Russian]. = Soviet Math. Dokl. **4**, 654–658 (1963).

134. Talalyan, A. A.: Representation of measurable functions by series. Uspehi Mat. Nauk **15**, No. 5, 77–141 (1960) [Russian].

135. Talalyan, A. A.: Complete systems of unconditional convergence in the weak sense. Izv. Akad. Nauk SSSR Ser. Mat. **28**, 713–720 (1964) [Russian].

136. Talalyan, A. A.: Systems of functions whose series represent in the metric $L_p[0,1]$ functions of the space $L_q[0,1]$, $1 \le p \le q$. Izv. Akad. Nauk Armjan. SSR Ser. Mat. **3**, 327–357 (1968) [Russian].

137. Talalyan, A. A.: On series universal with respect to rearrangements. Izv. Akad. Nauk SSSR Ser. Mat. **24**, 567–604 (1960) [Russian].

138. Talalyan, A. A.: Questions of representation and uniqueness in the theory of orthogonal series. In the collection Itogi Nauki: Matem. Analiz. Moscow, 1971, 5–64 [Russian].

139. Talalyan, A. A., Arutyunyan, F. G.: On divergence to $+\infty$ of series in the Haar system. Mat. Sb. **66**, 240–247 (1965) [Russian].

140. Talalyan, F. A.: Absolute and unconditional convergence. Izv. Akad. Nauk Armjan. SSR Ser. Mat. **5**, 108–137 (1970) [Russian].

141. Tandori, K.: Über die orthogonalen Funktionen, X. (Unbedingte Konvergenz). Acta Sci. Math. **23**, 185–221 (1962).

142. Tandori, K.: Über die Konvergenz der Orthogonalreihen. ibid. **24**, 139–151 (1963); II ibid. **25**, 219–232 (1964); III Publ. Math. Debrecen **12**, 127–157 (1965).

143. Tandori, K.: Über die Divergenz der Walshschen Reihen. Acta Sci. Math. **27**, 261–263 (1966).

144. Tandori, K.: Abschätzungen vom Menchoff-Rademacherschen Typ für die Summen von orthogonalen Funktionen. Studia Sci. Math. Hungar. **3**, 325–336 (1968).

145. Tandori, K.: Über die Reflexivität gewisser Banachräume. Acta Sci. Math. **30**, 39–42 (1969).

146. Tandori, K.: Ergänzung zu einem Satz von S. Kaczmarz. Acta Sci. Math. **28**, 147–153 (1967).

147. Tandori, K.: Über die unbedingte Konvergenz der Orthogonalreihen. Acta Sci. Math. **32**, 11–40 (1970).

148. Telyakovskiĭ, S. A.: Conditions for the integrability of trigonometric series, and their application to the study of linear methods of summation of Fourier series. Izv. Akad. Nauk SSSR Ser. Mat. **28**, 1209–1236 (1964) [Russian].

149. Tikhomirov, V. M.: Diameters of sets in function spaces and the theory of best approximations. Uspehi Mat. Nauk **15**, No. 3, 81–120 (1960) [Russian].

150. Tsereteli, O. D. (Cereteli): The unconditional convergence of orthogonal series and metric properties of functions. Sakharth. SSR Mecn. Akad. Moambe (formerly Soobšč. Akad. Nauk Gruzin. SSR) **59**, 21–24 (1970) [Russian].

151. Tsereteli, O. D. (Cereteli): The unconditional convergence of Fourier series in complete orthonormal systems. Sakharth. SSR Mecn. Akad. Moambe (formerly Soobšč. Akad. Nauk Gruzin. SSR) **62**, 13–16 (1971) [Russian].

152. Ulyanov, P. L.: Divergent Fourier series of class L^p, $p \geq 2$. Dokl. Akad. Nauk SSSR **137**, 786–789 (1961) [Russian]. = Soviet Math. Dokl. **2**, 350–354 (1961).

153. Ulyanov, P. L.: Divergent series obtained using the Haar system and using bases. Dokl. Akad. Nauk SSSR **138**, 556–559 (1961) [Russian]. = Soviet Math. Dokl. **2**, 679–682 (1961).

154. Ulyanov, P. L.: Divergent Fourier series. Uspehi Mat. Nauk **16**, No. 3, 61–142 (1961) [Russian].

155. Ulyanov, P. L.: Series with respect to the Haar system with monotone coefficients. Izv. Akad. Nauk SSSR Ser. Mat. **28**, 925–950 (1964) [Russian].

156. Ulyanov, P. L.: On Weyl multipliers for unconditional convergence. Mat. Sb. **60**, 39–62 (1963) [Russian].

157. Ulyanov, P. L.: Series in the Haar system. Mat. Sb. **63**, 356–391 (1964) [Russian].

158. Ulyanov, P. L.: Solved and unsolved problems in the theory of trigonometric and orthogonal series. Uspehi Mat. Nauk **19**, No. 1, 3–69 (1964) [Russian].

159. Ulyanov, P. L.: Some questions in the theory of orthogonal and biorthogonal series. Izv. Akad. Nauk Azerbaĭdžan. SSR, 11–13 (1965) [Russian].

160. Ulyanov, P. L.: Representation of functions by series and the classes $\phi(L)$. Uspehi Mat. Nauk **17**, 3–52 (1972) [Russian].

161. Ulyanov, P. L.: Unconditional convergence and summability. Izv. Akad. Nauk SSSR Ser. Mat. **22**, 811–840 (1958) [Russian].

162. Veselov, V. M.: Systems of convergence and divergence in the spaces C and B. Mat. Zametki **13**, 911–922 (1973) [Russian]. = Math. Notes **13**, 542–548 (1973).

163. Vinogradova, I. A., Skvortsov, V. A.: Generalized integrals and Fourier series. In the collection Itogi Nauki: Matem. Analiz. Moscow, 1971, 65–107 [Russian].

164. Weyl, H.: Über die Konvergenz von Reihen, die nach Orthogonalfunktionen fortschreiten. Math. Ann. **67**, 225–245 (1909).

165. Wiener, N., Paley, R.: Fourier transforms in the complex domain. New York: Amer. Math. Soc. Coll. Publ. XIX, 1934, 1–184. Russian translation: Moscow, 1964.

166. Wojtyński, W.: On conditional bases in non-nuclear Fréchet spaces. Studia Math. **35**, 77–96 (1970).

167. Yung-min Chen: Theorems of asymptotic approximation. Math. Ann. **140**, 360–407 (1960).

168. Zahorski, Z.: Une série Fourier permutée d'une fonction de classe L^2 divergente presque partout. C. R. Acad. Sci. Paris **251**, 501–503 (1960).

169. Zahorski, Z.: Sur les ensembles des points de divergence de certaines intégrales singulières. Ann. Soc. Polon. Math. **19**, 66–105 (1947).

170. Ziza, O. A.: On the summation of orthogonal series by Euler methods. Mat. Sb. **66**, 354–377 (1965) [Russian].

171. Zygmund, A.: Trigonometric series. 2 vols. 2 ed. Cambridge: Cambridge U. Pr. 1959. Russian translation: Moscow, 1965.

Subject Index

Ergebnisse der Mathematik und ihrer Grenzgebiete